JOURNAL OF APPLIED LOGICS - IFCOLOG JOURNAL OF LOGICS AND THEIR APPLICATIONS

Volume 11, Number 5

October 2024

Disclaimer

Statements of fact and opinion in the articles in Journal of Applied Logics - IfCoLog Journal of Logics and their Applications (JALs-FLAP) are those of the respective authors and contributors and not of the JALs-FLAP. Neither College Publications nor the JALs-FLAP make any representation, express or implied, in respect of the accuracy of the material in this journal and cannot accept any legal responsibility or liability for any errors or omissions that may be made. The reader should make his/her own evaluation as to the appropriateness or otherwise of any experimental technique described.

ISBN 978-1-84890-469-9
ISSN (E) 2631-9829
ISSN (P) 2631-9810

College Publications
Scientific Director: Dov Gabbay
Managing Director: Jane Spurr

http://www.collegepublications.co.uk

SCOPE AND SUBMISSIONS

This journal considers submission in all areas of pure and applied logic, including:

pure logical systems
proof theory
constructive logic
categorical logic
modal and temporal logic
model theory
recursion theory
type theory
nominal theory
nonclassical logics
nonmonotonic logic
numerical and uncertainty reasoning
logic and AI
foundations of logic programming
belief change/revision
systems of knowledge and belief
logics and semantics of programming
specification and verification
agent theory
databases

dynamic logic
quantum logic
algebraic logic
logic and cognition
probabilistic logic
logic and networks
neuro-logical systems
complexity
argumentation theory
logic and computation
logic and language
logic engineering
knowledge-based systems
automated reasoning
knowledge representation
logic in hardware and VLSI
natural language
concurrent computation
planning

This journal will also consider papers on the application of logic in other subject areas: philosophy, cognitive science, physics etc. provided they have some formal content.

Submissions should be sent to Jane Spurr (jane@janespurr.net) as a pdf file, preferably compiled in LaTeX using the IFCoLog class file.

CONTENTS

ARTICLES

vii

Special Issue on Frontiers of Logic and Computation in Iran

ARSHAM BORUMAND SAEID
Shahid Bahonar University of Kerman, Kerman, Iran.

RAJAB ALI BORZOOEI
Shahid Beheshti University, Tehran, Iran.

MOHAMMAD MEHDI ZAHEDI
Graduate University of Advanced Technology, Kerman, Iran.

The reach of logic has expanded significantly in recent years, permeating an ever-growing number of natural and social science disciplines. From the traditional applications in Mathematics, physics, and computer science, logic has now found applications in fields as diverse as philosophy, cognitive science, and linguistics. Moreover, its influence has extended into virtually all aspects of information technology, from software engineering and hardware design to programming and artificial intelligence. This idea has led to the emergence of a new interdisciplinary area at the intersection of logic, artificial intelligence, cognitive science, and theoretical computing. Indeed, the synergies between these fields have become increasingly pronounced, with each discipline informing and inspiring the others. Advances in logic have enabled breakthroughs in artificial intelligence, as the rigorous formalisms of logic provide a foundation for building intelligent systems. The challenges faced in developing robust and flexible AI have, in turn, pushed the boundaries of logical reasoning, leading to the development of novel logics and reasoning frameworks. Similarly, the study of cognition and the human mind, a core focus of cognitive science, has both drawn upon and contributed to the evolution of logical concepts and tools.

This dynamic interplay has created a rapidly evolving landscape, where the traditional boundaries between disciplines are blurring, and new avenues for exploration and discovery are constantly emerging. It is against this backdrop that Iranian scholars have made significant contributions, pioneering new directions in logic and its applications across a wide range of domains.

Iranian researchers have been at the forefront of this exciting development, making substantial contributions to the field through their research. In recent years, they have made great strides in promoting logical studies and have achieved remarkable success across a broad spectrum of topics, including classical and non-classical logics, algebraic logic, modal and temporal logic, probabilistic logic, aggregation functions and fuzzy implication, knowledge-based systems and knowledge representation, and automated reasoning.

The aim of this special issue is to provide a platform for Iranian scholars to share their novel ideas, original research achievements, and practical experiences with the international community. Submissions are invited on advancements in classical and non-classical logics, developments in algebraic logic, new frontiers in modal and temporal logic, innovations in probabilistic and fuzzy logic, cutting-edge research in knowledge representation and reasoning, breakthroughs in automated theorem proving and decision procedures, and innovative applications of logic in artificial intelligence, computer science, and cognitive science. This special issue offers a unique opportunity for the global research community to learn about the latest progress in logic and computation from leading Iranian scholars, fostering cross-cultural exchange and collaboration.

Topics include, but are certainly not limited to:

- Non-classical logic and Non-monotonic logic

- Algebraic logic(MV-algebra, Bl-algebra, Heyting algebra,...)

- Temporal logic, modal logic and intuitionistic logic

- Probabilistic logic and Fuzzy logic

- Aggregation function and Fuzzy implication

- Approximation reasoning and Automated reasoning

- Soft Computing and Granular Computing

Guest Editors
Arsham Borumand Saeid
Rajab Ali Borzooei
Mohammad Mehdi Zahedi

    Received August 2024

SOME TYPES OF σ-FILTERS IN BL-ALGEBRAS

ZAHRA PARVIZI

Department of Mathematics, Shahrekord Branch, Islamic Azad University, Shahrekord, Iran

SOMAYEH MOTAMED

Department of Mathematics, Bandar Abbas Branch, Islamic Azad University, Bandar Abbas, Iran

FARHAD KHAKSAR HAGHANI

Department of Mathematics, Shahrekord Branch, Islamic Azad University, Shahrekord, Iran

Abstract

In this paper, we define σ-normal filters, σ-fantastic filters, σ-obstinate filters, σ-integral filters, σ-semi-integral filters in BL-algebras, then we state and prove some theorems to characterize these concepts. In this paper, we begin an algebraic investigation of new types of σ-filters in BL-algebras; which contains definitions, and their relationship. Finally, equivalent conditions are provided for easier study of these types of filters in BL-algebras and some important classes of BL-algebras and their connections are presented.

1 Introduction

Since Hájek published his book [8], fuzzy logic originally introduced by Zadeh in [17] has received a strong impetus into the direction of mathematical logic; to date mathematical fuzzy logic has been analyzed in several hundreds of scientific works. One of the fundamental discoveries of Hájek are Basic fuzzy logic and BL-algebras; they are in the same relation with each other than classical logic and Boolean algebras. Since filters played an important role in the study of these algebras, the types of filters in these algebras were defined and studied over several years. To see the results, you can refer to the articles in this field [2, 3, 4, 9, 12, 15], ... After that Najmi Dolat Abadi and Moghaderi in 2019 [13], have introduced the notions of σ-filter, σ-invariant and σ-primary for filters of BL-algebras and studied their characteristics and we continue their work. Based on these results, in this paper some types of σ-filters on BL-algebras have been studied and some important results have been obtained. Since BL-algebras are important algebras among logical algebraic structures, our motivation was also to further investigate BL-algebras with the help of various σ-filters in the continuation of the work of Najmi Dolat Abadi and J. Moghaderi, [13]. We begin in this paper an algebraic investigation of new types of σ-filters in BL-algebras; it contains definitions, basic properties and the relationship between them. This paper aims to analyze the structure of BL-algebras by new types of σ-filters. Since filters play a very important role in examining different structures of BL-algebras, in this article we tried to introduce and study a new σ-filter for studying BL-algebras as much as possible. Our motivation was to study σ-filters as possible in BL-algebras, to be able to obtain new relationships between σ-filters as well as different structures of BL-algebras, including MV-algebras, Boolean algebras, etc.

2 Preliminaries

Definition 2.1. [8] A BL-algebra is an algebra $(L, \wedge, \vee, \odot, \rightarrow, 0, 1)$ with four binary operations $\vee, \wedge, \odot, \rightarrow$ and two constants 0, 1 such that

(BL1) $(L, \wedge, \vee, 0, 1)$ is a bounded lattice;

(BL2) $(L, \odot, 1)$ is a commutative monoid;

(BL3) $\odot$ and $\rightarrow$ form an adjoint pair, i.e. $c \leq a \rightarrow b$, if and only if $a \odot c \leq b$, for all $a, b, c \in L$;

(BL4) $a \wedge b = a \odot (a \rightarrow b)$;

(BL5) $(a \rightarrow b) \vee (b \rightarrow a) = 1$.

Recall that for a BL-algebra L, $x^* = x \rightarrow 0$, for all $x \in L$.

A BL-algebra L is called:

- an MV-algebra, if for all $x \in L$, $x^{**} = x$, where $x^* = x \rightarrow 0$, [8].
- a Gödel algebra, if for all $x \in L$, $x^2 = x$, where $x^2 = x \odot x$, [5].
- a special BL-algebra, if for all $0 \neq a, b \in L$, $(a \rightarrow b)^* = (b \rightarrow a)^*$, [11].
- an integral BL-algebra, if for any $x, y \in L$, $x \odot y = 0$ implies $x = 0$ or $y = 0$, [4].

- a semi-integral BL-algebra, if for any $x, y \in L$, $x \odot y = 0$ implies $x = 0$ or $y^n = 0$, for some $n \in \mathbb{N}$, [12].

Definition 2.2. [4] L is called a BL-algebra with Gödel negation, if $\{a \in L : a \rightarrow 0 = 0\} = L \smallsetminus \{0\}$.

Theorem 2.3. *[5, 6, 8, 16] In any BL-algebra L, the following properties hold for all $x, y, z \in L$:*

(BL6) $x \le y$, *if and only if* $x \to y = 1$;

(BL7) $x \to (y \to z) = y \to (x \to z) = (x \odot y) \to z$;

(BL8) *if* $x \le y$, *then* $y \to z \le x \to z$ *and* $z \to x \le z \to y$;

(BL9) $1 \to x = x$, $x \to x = 1$, $x \le y \to x$, $x \to 1 = 1$, $0 \to x = 1$;

(BL10) $x \odot x^* = 0$ *and* $0^* = 1$;

(BL11) $x \to y \le (y \to z) \to (x \to z)$ *and* $x \to y \le (z \to x) \to (z \to y)$;

(BL12) $x \to y \le (z \odot x) \to (z \odot y)$;

(BL13) $x \to (y \wedge z) = (x \to y) \wedge (x \to z)$.

Let F be a non-empty subset of a BL-algebra L and $x, y \in L$. Then the following conditions are equivalent:

(*i*) F is a filter of L;

(*ii*) If $x, y \in F$ imply $x \odot y \in F$ and if $x \in F$ and $x \le y$, then $y \in F$;

(*iii*) $1 \in F$ and if $x, x \to y \in F$ imply $y \in F$, [8].

$D(X) = \{x \in L : x^{**} \in X\}$, where X is a subset of L. $D(X)$ is a filter, when X is a filter of L, [1].

A Filter F of L is proper if $F \ne L$; that is, $0 \notin F$. A proper filter is maximal if it isn't contained in any other proper filter.

Definition 2.4. A non-empty subset F of a BL-algebra L is called:

• a Boolean filter of L, if F is a filter and for all $x \in L$, $x \vee x^* \in F$, [8].

• a positive implicative of L, if $1 \in F$ and for all $x, y, z \in L$, if $x \to ((y \to z) \to y) \in F$, $x \in F$ imply $y \in F$, [9].

• a fantastic filter of L, if $1 \in F$ and for all $x, y, z \in L$, if $z \to (y \to x) \in F$ and $z \in F$ imply $((x \to y) \to y) \to x \in F$, [9].

• an obstinate filter of L, if F is a proper filter of L and for all $x, y, z \in L$, if $x, y \notin F$ imply $x \to y \in F$ and $y \to x \in F$, [3].

• an integral filter of L, if F is a proper filter and for all $x, y \in L$, if $(x \odot y)^* \in F$ imply $x^* \in F$ or $y^* \in F$, [4].

• a semi-integral filter of L, if F is a proper filter and for all $x, y \in L$, if $(x \odot y)^* \in F$, then there exists $n \in \mathbb{N}$ such that $x^* \in F$ or $(y^n)^* \in F$, [12].

• a primary filter of L, if F is a proper filter of L and for all $x, y \in L$, if $(x * y)^- \in F$ implies $(x^n)^- \in F$ or $(y^n)^- \in F$, $\exists n \in N$, [18].

Definition 2.5. [7, 8, 16] Let A and B be two BL-algebras. A map $f : A \to B$ defined on A, is called a BL-homomorphism, if for all $x, y \in A$, $f(x \to y) = f(x) \to f(y)$, $f(x \odot y) = f(x) \odot f(y)$, and $f(0_A) = 0_B$.

Definition 2.6. [13], [14] Let $\sigma : L \to L$ be a BL-homomorphism. Then a filter F of L is called

(i) a σ-filter, if $\sigma(F) \subseteq F$;

(ii) a σ-positive implicative filter, if $\sigma(x) \to ((\sigma(y) \to \sigma(z)) \to \sigma(y)) \in F$ and $x \in \sigma(F)$ imply $y \in \sigma(F)$, for all $x, y, z \in L$;

(iii) a σ-strongly primary, if F is proper and for $x, y \in L$, $(\sigma(x) \odot \sigma(y))^* \in F$ implies $(x^n)^* \in \sigma(F)$ or $(y^n)^* \in \sigma(F)$, for some $n \in \mathbb{N}$.

• The relation $\sim_F$ defined on a BL-algebra L, by $(x, y) \in \sim_F$, if and only if $x \to y \in F$ and $y \to x \in F$ is a congruence relation on L, for any proper filter F of a BL-algebra L. The quotient algebra $L/\sim_F$ denoted by L/F becomes a BL-algebra in a natural way, with the operations induced from those of L. So the order relation on L/F is given by $x/F \leq y/F$, if and only if $x \to y \in F$. Hence $x/F = 1/F$, if and only if $x \in F$ and $x/F = 0/F$, if and only if $x^* \in F$, [8].

• Let L be a BL-algebra, σ be a BL-endomorphism of L and F be a proper σ-filter of L. Then the function mapping $\overline{\sigma} : L/F \to L/F$, by $\overline{\sigma}[x] = [\sigma(x)]$, is well defined, [13].

Theorem 2.7. *[13] (i) Let X be a non-empty subset of L. Then $\sigma(D(X)) = D(\sigma(X))$ when σ is an isomorphism.*

(ii) Let F be a σ-filter of L. Then $D(F)$ is a σ-filter of L.

Lemma 2.8. *[9] In any BL-algebra L, the following are equivalent, for all $x, y, z \in L$:*

(a) $((x \to y) \to y) \to x = y \to x$;

(b) *If $x \to z \leq y \to z$, $z \leq x$, then $y \leq x$.*

(c) *If $x \to z \leq y \to z$, $z \leq x, y$, then $y \leq x$;*

(d) *If $y \leq x$, then $(x \to y) \to y \leq x$;*

(e) *L is an MV-algebra.*

Lemma 2.9. *[14] Let P be a σ-filter σ-strongly primary filter of a BL-algebra L and $\overline{\sigma}([x]) \odot \overline{\sigma}([y]) = [0]$, for $[x], [y] \in L/P$. Then $[x]$ or $[y]$ is a nilpotent element of L/P.*

Throughout this paper, it is assumed that $(L, \wedge, \vee, \odot, \to, 0, 1)$ is a BL-algebra and $\sigma : L \to L$ is a BL-homomorphism.

3 σ-Fantastic Filters in BL-algebras

In this section, we introduce the notion of σ-fantastic filters and investigate some properties of them.

Definition 3.1. A filter F of L is called a σ-fantastic filter, if $\sigma(z) \to (\sigma(y) \to \sigma(x)) \in F$ and $z \in \sigma(F)$ imply $((x \to y) \to y) \to x \in \sigma(F)$, for all $x, y, z \in L$.

Example 3.2. (i) Let $L = \{0, a, b, 1\}$, where $0 < a, b < 1$. The operations $\odot$ and $\to$ are defined as follows:

$\odot$	0	a	b	1
0	0	0	0	0
a	0	a	0	a
b	0	0	b	b
1	0	a	b	1

$\to$	0	a	b	1
0	1	1	1	1
a	b	1	b	1
b	a	a	1	1
1	0	a	b	1

Then $(L, \wedge, \vee, \odot, \to, 0, 1)$ is a BL-algebra [10] and $\sigma : L \to L$ is a BL-homomorphism, when $\sigma(a) = b$, $\sigma(b) = a$. It is easy to check that $F = \{a, 1\}$ and $G = \{b, 1\}$ are σ-fantastic filters of L.

(ii) Let $L = \{0, a, b, c, 1\}$ where $0 < c < a, b < 1$. The operations $\odot$ and $\to$ are defined as follows:

$\odot$	0	a	b	c	1
0	0	0	0	0	0
a	0	a	c	c	a
b	0	c	b	c	b
c	0	c	c	c	c
1	0	a	b	c	1

$\to$	0	a	b	c	1
0	1	1	1	1	1
a	0	1	b	b	1
b	0	a	1	a	1
c	0	1	1	1	1
1	0	a	b	c	1

Then $(L, \wedge, \vee, \odot, \to, 0, 1)$ is a BL-algebra [7] and σ is a BL-homomorphism when $\sigma(a) = b$, $\sigma(b) = a$, $\sigma(c) = c$. Clearly, $H = \{a, 1\}$ isn't σ-fantastic filter, since $\sigma(b) \to (\sigma(0) \to \sigma(a)) = 1 \in H$ and $b \in \sigma(H)$ but $((a \to 0) \to 0) \to a = a \notin \sigma(H)$.

(iii) Let $L = \{0, a, b, c, 1\}$ where $0 < a < c, b < 1$. The operations $\odot$ and $\to$ are defined as follows:

$\odot$	0	a	b	c	1
0	0	0	0	0	0
a	0	a	a	a	a
b	0	a	b	a	b
c	0	a	a	c	c
1	0	a	b	c	1

$\to$	0	a	b	c	1
0	1	1	1	1	1
a	0	1	1	1	1
b	0	c	1	c	1
c	0	b	b	1	1
1	0	a	b	c	1

Then $(L, \wedge, \vee, \odot, \to, 0, 1)$ is a BL-algebra [10]. Take BL-homomorphism σ where $\sigma(a) = b$, $\sigma(b) = 1$ and $\sigma(c) = b$. It is clear that $F = \{c, 1\}$ is a σ-fantastic filter of L, while F isn't a fantastic filter. Since $1 \to (0 \to a) = 1 \in F$ and $1 \in F$ but $((a \to 0) \to 0) \to a = a \notin F$.

(*iv*) Let $L = \{0, a, b, 1\}$, where $0 < a < b < 1$. The operations $\odot$ and $\to$ are defined as follows:

$\odot$	0	a	b	1
0	0	0	0	0
a	0	0	a	a
b	0	a	b	b
1	0	a	b	1

$\to$	0	a	b	1
0	1	1	1	1
a	a	1	1	1
b	0	a	1	1
1	0	a	b	1

Then $(L, \wedge, \vee, \odot, \to, 0, 1)$ is a BL-algebra [10]. Consider $\sigma(a) = a$ and $\sigma(b) = 1$. It is easy to verify that $F = \{b, 1\}$ is a fantastic filter, while F isn't a σ-fantastic filter of L, since $\sigma(1) \to (\sigma(a) \to \sigma(b)) = 1 \in F$ and $1 \in \sigma(F)$ but $((b \to a) \to a) \to b = b \notin \sigma(F)$.

Theorem 3.3. *Let F be a σ-fantastic filter of L. Then $\sigma(y) \to \sigma(x) \in F$ implies $((x \to y) \to y) \to x \in \sigma(F)$, for all $x, y \in L$.*

Proof. Let F be a σ-fantastic filter and $\sigma(y) \to \sigma(x) \in F$, for all $x, y \in L$. Hence $\sigma(1) \to (\sigma(y) \to \sigma(x)) = \sigma(y) \to \sigma(x) \in F$, and since $1 \in \sigma(F)$ then by hypothesis $((x \to y) \to y) \to x \in \sigma(F)$. $\square$

By adding a condition, we proved the converse of Theorem 3.3:

Theorem 3.4. *Let F be a σ-filter of L and $\sigma(y) \to \sigma(x) \in F$ implies $((x \to y) \to y) \to x \in \sigma(F)$, for all $x, y \in L$. Then F is a σ-fantastic filter of L.*

Proof. Let F be a σ-filter and $\sigma(z) \to (\sigma(y) \to \sigma(x)) \in F$ and $z \in \sigma(F)$. Then $\sigma(z) \in F$, since F is a σ-filter of L. Then we have $\sigma(y) \to \sigma(x) \in F$. By hypothesis, we get that $((x \to y) \to y) \to x \in \sigma(F)$. Therefore F is a σ-fantastic filter of L. $\square$

The following example, shows that the condition in Theorem 3.4 is necessary.

Example 3.5. Consider Example 3.2(*ii*), when $\sigma(a) = b$, $\sigma(b) = a$ and $\sigma(c) = c$. Clearly, $F = \{a, 1\}$ isn't a σ-filter and $\sigma(y) \to \sigma(x) \in F$ implies $((x \to y) \to y) \to x \in \sigma(F)$, for all $x, y \in L$, while F isn't a σ-fantastic filter, since $\sigma(b) \to (\sigma(0) \to \sigma(a)) = 1 \in F$ and $b \in \sigma(F)$ but $((a \to 0) \to 0) \to a = a \notin \sigma(F)$.

Using Theorems 3.3 and 3.4, we can conclude that:

Corollary 3.6. *Let F be a σ-filter of L. Then F is a σ-fantastic filter of L if and only if $\sigma(y) \to \sigma(x) \in F$ implies $((x \to y) \to y) \to x \in \sigma(F)$, for all $x, y \in L$.*

600

Theorem 3.7. *Let σ be a BL-isomorphism on L and F be a σ-fantastic σ-filter of L. Then $D(F)$ is a σ-fantastic filter of L.*

Proof. Let $\sigma(y) \to \sigma(x) \in D(F)$, for $x, y \in L$. Then $\sigma(y^{**}) \to \sigma(x^{**}) \in F$. Hence by Theorem 3.3, $((x^{**} \to y^{**}) \to y^{**}) \to x^{**} \in \sigma(F)$. Thus we get that $((x \to y) \to y) \to x \in D(\sigma(F))$. By using Theorem 2.7(i), $((x \to y) \to y) \to x \in \sigma(D(F))$. As according to Theorem 2.7(ii), $D(F)$ is a σ-filter, by Theorem 3.4, $D(F)$ is a σ-fantastic filter of L. $\qquad\square$

Theorem 3.8. *Let $\{1\}$ be a σ-fantastic filter of L. Then $(x \to y) \to y = (y \to x) \to x$, for all $x, y \in L$.*

Proof. Assume that $\{1\}$ is a σ-fantastic filter and $a = (y \to x) \to x$, for $x, y \in L$. Then $\sigma(y) \to \sigma(a) = \sigma(y) \to ((\sigma(y) \to \sigma(x)) \to \sigma(x)) = (\sigma(y) \to \sigma(x)) \to (\sigma(y) \to \sigma(x)) = 1$. So by hypothesis and Theorem 3.3, $((a \to y) \to y) \to a = 1$. So $(a \to y) \to y \le a$. Since $x \le a$, we have $a \to y \le x \to y$ and so $(x \to y) \to y \le (a \to y) \to y$. Hence $1 = ((a \to y) \to y) \to a \le ((x \to y) \to y) \to a$. Then $((x \to y) \to y) \to ((y \to x) \to x) = 1$, i.e. $(x \to y) \to y \le (y \to x) \to x$ and simillary, $(y \to x) \to x \le (x \to y) \to y$, therefore $(x \to y) \to y = (y \to x) \to x$, for $x, y \in L$. $\qquad\square$

Corollary 3.9. *Let $\{1\}$ be a σ-fantastic filter of L. Then for all $x, y, z \in L$:*
(1) $((x \to y) \to y) \to x = y \to x$;
(2) *If $x \to z \le y \to z$, $z \le x$, then $y \le x$.*
(3) *If $x \to z \le y \to z$, $z \le x, y$, then $y \le x$;*
(4) *If $y \le x$, then $(x \to y) \to y \le x$;*
(5) *L is an MV-algebra.*

Proof. Using Theorem 3.8 and Lemma 2.8, the proof is clear. $\qquad\square$

If we put $y = 0$ in Theorem 3.3, we get:

Corollary 3.10. *Let F be a σ-fantastic filter of L. Then $x^{**} \to x \in \sigma(F)$, for all $x \in L$.*

The following example shows that the converse of the Corollary 3.10, isn't true in general.

Example 3.11. Consider Example 3.2(i), when $\sigma(a) = 1$ and $\sigma(b) = 0$. It is clear that $x^{**} \to x \in \sigma(F)$, for all $x \in L$, while $F = \{a, 1\}$ isn't σ-fantastic filter, since $\sigma(b) \to \sigma(0) \in F$ but $((0 \to b) \to b) \to 0 = a \notin \sigma(F)$.

Lemma 3.12. *Let F be a σ-fantastic σ-filter of L. Then $\sigma(x) \to \sigma(u) \in F$ and $\sigma(y) \to \sigma(u) \in F$ imply $((\sigma(x) \to \sigma(y)) \to \sigma(y)) \to \sigma(u) \in F$, for all $x, y, u \in L$.*

Proof. If $\sigma(x) \to \sigma(u)$, $\sigma(y) \to \sigma(u) \in F$, then by (BL11),

$$\sigma(x) \to \sigma(u) \leq (\sigma(u) \to 0) \to (\sigma(x) \to 0)$$

and

$$\sigma(y) \to \sigma(u) \leq (\sigma(u) \to 0) \to (\sigma(y) \to 0),$$

then we have

$$(\sigma(u) \to 0) \to (\sigma(x) \to 0) \in F \text{ and } (\sigma(u) \to 0) \to (\sigma(y) \to 0) \in F.$$

Thus by the property of filters

$$((\sigma(u) \to 0) \to (\sigma(x) \to 0)) \wedge ((\sigma(u) \to 0) \to (\sigma(y) \to 0)) \in F.$$

We have

$$((\sigma(u) \to 0) \to (\sigma(x) \to 0)) \wedge ((\sigma(u) \to 0) \to (\sigma(y) \to 0)) \overset{(BL13)}{=}$$

$$(\sigma(u) \to 0) \to ((\sigma(x) \to 0) \wedge (\sigma(y) \to 0)) \overset{(BL4)}{=}$$

$$(\sigma(u) \to 0) \to ((\sigma(y) \to 0) \odot ((\sigma(y) \to 0) \to (\sigma(x) \to 0))) \overset{(BL7)}{=}$$

$$(\sigma(u) \to 0) \to ((\sigma(y) \to 0) \odot (\sigma(x) \to ((\sigma(y) \to 0) \to 0))) \in F,$$

it follows from

$$((\sigma(u) \to 0) \to ((\sigma(y) \to 0) \odot (\sigma(x) \to ((\sigma(y) \to 0) \to 0)))) \to$$

$$((\sigma(u) \to 0) \to ((\sigma(y) \to 0) \odot (\sigma(x) \to \sigma(y)))) \overset{(BL11)}{\geq}$$

$$((\sigma(y) \to 0) \odot (\sigma(x) \to ((\sigma(y) \to 0) \to 0))) \to$$

$$((\sigma(y) \to 0) \odot (\sigma(x) \to \sigma(y))) \overset{(BL12)}{\geq}$$

$$(\sigma(x) \to ((\sigma(y) \to 0) \to 0)) \to (\sigma(x) \to \sigma(y)) \overset{(BL11)}{\geq}$$

$$((\sigma(y) \to 0) \to 0) \to \sigma(y) \in F, \text{ (Corollary 3.10)},$$

that

$$(\sigma(u) \to 0) \to ((\sigma(y) \to 0) \odot (\sigma(x) \to \sigma(y))) \in F.$$

Morever, from

$$(\sigma(u) \to 0) \to ((\sigma(y) \to 0) \odot (\sigma(x) \to \sigma(y))) \overset{\text{(BL11)}}{\leq}$$

$$(((\sigma(y) \to 0) \odot (\sigma(x) \to \sigma(y))) \to 0) \to ((\sigma(u) \to 0) \to 0) \overset{\text{(BL7)}}{=}$$

$$((\sigma(x) \to \sigma(y)) \to ((\sigma(y) \to 0) \to 0)) \to ((\sigma(u) \to 0) \to 0),$$

we have

$$((\sigma(x) \to \sigma(y)) \to ((\sigma(y) \to 0) \to 0)) \to ((\sigma(u) \to 0) \to 0) \in F.$$

Since

$$((\sigma(x) \to \sigma(y) \to ((\sigma(y) \to 0) \to 0)) \to ((\sigma(u) \to 0) \to 0) \in F$$

and

$$(((\sigma(x) \to \sigma(y)) \to ((\sigma(y) \to 0) \to 0)) \to ((\sigma(u) \to 0) \to 0)) \to$$

$$(((\sigma(x) \to \sigma(y)) \to \sigma(y)) \to ((\sigma(u) \to 0) \to 0)) \overset{\text{(BL11)}}{\geq}$$

$$((\sigma(x) \to \sigma(y)) \to \sigma(y)) \to ((\sigma(x) \to \sigma(y)) \to ((\sigma(y) \to 0) \to 0)) \overset{\text{(BL11)}}{\geq}$$

$$\sigma(y) \to ((\sigma(y) \to 0) \to 0) \quad = 1 \in F,$$

we get that

$$((\sigma(x) \to \sigma(y)) \to \sigma(y)) \to ((\sigma(u) \to 0) \to 0) \in F.$$

It follows from

$$(((\sigma(x) \to \sigma(y)) \to \sigma(y)) \to ((\sigma(u) \to 0) \to 0)) \to$$

$$(((\sigma(x) \to \sigma(y)) \to \sigma(y)) \to \sigma(u)) \overset{\text{(BL11)}}{\geq}$$

$$((\sigma(u) \to 0) \to 0) \to \sigma(u) \quad \in F, \qquad \text{(By Corollary 3.10)},$$

that

$$((\sigma(x) \to \sigma(y)) \to \sigma(y)) \to \sigma(u) \in F.$$

$\square$

Corollary 3.13. *Let F be a σ-fantastic filter and $\sigma(F)$ be a proper σ-filter of L. Then $L/\sigma(F)$ is an MV-algebra.*

Proof. Let F be a σ-fantastic filter. By Corollary 3.10, $x^{**} \to x \in \sigma(F)$, for all $x \in L$. So $((x/\sigma(F) \to 0/\sigma(F)) \to 0/\sigma(F)) \to x/\sigma(F) = 1/\sigma(F)$ in $L/\sigma(F)$. Hence $(x/\sigma(F))^{**} = x/\sigma(F)$ in $L/\sigma(F)$. Therefore $L/\sigma(F)$ is an MV-algebra. $\square$

Theorem 3.14. *Let F be a σ-fantastic filter of L. Then $\sigma(x^*) \in F$ implies $x^* \in \sigma(F)$, for all $x \in L$.*

Proof. Let $\sigma(x^*) \in F$. We know that $\sigma(x) \to 0 \in F$. Since F is a σ-fantastic filter, then we get that $((0 \to x) \to x) \to 0 = x^* \in \sigma(F)$. $\qquad\square$

The following example shows that the converse of Theorem 3.14 isn't true in general.

Example 3.15. Consider Example 3.2(ii), when $\sigma(a) = b$, $\sigma(b) = a$ and $\sigma(c) = c$. Consider $F = \{a, 1\}$. It is easy to see that $\sigma(x^*) \in F$ implies $x^* \in \sigma(F)$, for any $x \in L$, while F isn't σ-fantastic filter, since $\sigma(b) \to (\sigma(0) \to \sigma(a)) = a \to (0 \to b) = 1 \in F$, $b \in \sigma(F)$ but $((a \to 0) \to 0) \to a = a \notin \sigma(F)$.

Lemma 3.16. *Let F be a σ-fantastic of L and $\sigma(F)$ be a proper filter of L with Gödel negation. Then $\sigma(F) = L - \{0\}$.*

Proof. Let $0 \neq x \in L$ and F be a σ-fantastic filter of L. Then by Corollary 3.10, $((x \to 0) \to 0) \to x \in \sigma(F)$. Since L is a BL-algebra with Gödel negation, then $x \to 0 = 0$ and so we have $((x \to 0) \to 0) \to x = x \in \sigma(F)$. Hence $L - \{0\} \subseteq \sigma(F)$ and so $\sigma(F) = L - \{0\}$. $\qquad\square$

Theorem 3.17. *Let F be a σ-positive implicative σ-filter of L. Then F is a σ-fantastic filter of L.*

Proof. Let $x, y \in L$ and $\sigma(y) \to \sigma(x) \in F$. We show that $((x \to y) \to y) \to x \in \sigma(F)$, by Theorem 3.4. We have $\sigma(x) \leq ((\sigma(x) \to \sigma(y)) \to \sigma(y)) \to \sigma(x)$. By Theorem 2.3, we get $(((\sigma(x) \to \sigma(y)) \to \sigma(y)) \to \sigma(x)) \to \sigma(y) \leq \sigma(x) \to \sigma(y)$ and also we have $(((\sigma(x) \to \sigma(y)) \to \sigma(y)) \to \sigma(x)) \to \sigma(y)) \to (((\sigma(x) \to \sigma(y)) \to \sigma(y)) \to \sigma(x)) \geq (\sigma(x) \to \sigma(y)) \to (((\sigma(x) \to \sigma(y)) \to \sigma(y)) \to \sigma(x)) \geq ((\sigma(x) \to \sigma(y)) \to \sigma(y)) \to ((\sigma(x) \to \sigma(y)) \to \sigma(x)) \geq \sigma(y) \to \sigma(x)$. But since $\sigma(y) \to \sigma(x) \in F$, we get $(((\sigma(x) \to \sigma(y)) \to \sigma(y)) \to \sigma(x)) \to \sigma(y)) \to (((\sigma(x) \to \sigma(y)) \to \sigma(y)) \to \sigma(x)) \in F$. Since F is a σ-positive implicative filter of L, we conclude that $((x \to y) \to y) \to x \in \sigma(F)$. $\qquad\square$

The following example shows that the converse of Theorem 3.17, isn't true in general.

Example 3.18. Consider Example 3.2(iii), when $\sigma(a) = b$, $\sigma(b) = 1$ and $\sigma(c) = b$. It is easy to check that $F = \{c, 1\}$ is a σ-fantastic filter, while F isn't σ-positive implicative filter (since $\sigma(1) \to (((\sigma(a) \to \sigma(0)) \to \sigma(a)) = 1 \in F$ and $1 \in \sigma(F)$ but $a \notin \sigma(F))$.

Definition 3.19. Let F be a proper filter of L. F is called a σ-obstinate filter of L, if $\sigma(x), \sigma(y) \notin F$ imply $x \to y \in \sigma(F)$ and $y \to x \in \sigma(F)$, for all $x, y \in L$.

Example 3.20. (i) Consider Example 3.2(ii). $\sigma : L \to L$ is a BL-homomorphism when $\sigma(a) = b$, $\sigma(b) = 1$ and $\sigma(c) = b$. It is easy to check that $F = \{a, 1\}$, $G = \{b, 1\}$ are σ-obstinate filters of L.

(ii) Consider Example 3.2(ii), when $\sigma(a) = b$, $\sigma(b) = a$ and $\sigma(c) = c$. $H = \{b, 1\}$ isn't σ-obstinate filter, since $\sigma(c), \sigma(0) \notin H$ but $c \to 0 = 0 \notin \sigma(H)$.

Proposition 3.21. Let F be a σ-obstinate filter of L. If $\sigma(x) \notin F$, for $x \in L$, then $x^* \in \sigma(F)$.

Proof. Suppose that F is a σ-obstinate filter and $\sigma(x) \notin F$. Then $1 = 0 \to x \in \sigma(F)$ and $x^* = x \to 0 \in \sigma(F)$. $\qquad\qquad\square$

The following example shows that the converse of Proposition 3.21, isn't correct in general.

Example 3.22. Consider BL-algebra L in Example 3.2(iii), when $\sigma(a) = 0$, $\sigma(b) = 1$ and $\sigma(c) = 0$. Clearly, $\sigma(x) \notin F$, for any $x \in L$ implies $x^* \in \sigma(F)$, while F isn't a σ-obstinate filter of L, since $\sigma(a), \sigma(c) \notin F$ but $c \to a = b \notin \sigma(F)$.

Recall that a non-empty subset F of L is called upper set, if $x \le y$ and $x \in F$ implies $y \in F$, for $x, y \in L$.

Proposition 3.23. Let F be a proper filter of L and $\sigma(F)$ be an upper set of L. If $\sigma(x) \notin F$, $x^* \in \sigma(F)$, for all $x \in L$, then F is a σ-obstinate filter of L.

Proof. Let $\sigma(x), \sigma(y) \notin F$. We show that $x \to y \in \sigma(F)$ and $y \to x \in \sigma(F)$. By hypothesis $x^*, y^* \in \sigma(F)$. By Theorem 2.3, we have $x^* \le x \to y$ and $y^* \le y \to x$. Hence $x \to y \in \sigma(F)$ and $y \to x \in \sigma(F)$. $\qquad\qquad\square$

The following example, shows that the condition in Proposition 3.23 is necessary.

Example 3.24. Consider Example 3.2(iii), when $\sigma(a) = 0$, $\sigma(b) = 1$ and $\sigma(c) = 0$. It is clear that $\sigma(x) \notin F = \{c, 1\}$ implies $x^* \in \sigma(F)$, for all $x \in L$ and $\sigma(F)$ isn't upper set. And F isn't σ-obstinate filter, since $\sigma(a), \sigma(c) \notin F$ and $c \to a = b \notin \sigma(F)$.

By Proposition 3.23:

Corollary 3.25. Let F be a proper filter of L and $\sigma(F)$ be an upper set of L. If $\sigma(x) \notin F$ implies that $(x^*)^n \in \sigma(F)$, for some $n \in N$ and for all $x \in L$, then F is a σ-obstinate filter of L.

Using Proposition 3.21:

Theorem 3.26. *Let F be a proper filter of L and $\sigma(F)$ be an upper set of L. Then F is a σ-obstinate filter if and only if $\sigma(x) \in F$ or $x^* \in \sigma(F)$, for all $x \in L$.*

Proof. Assume that F is a σ-obstinate filter and $\sigma(x) \notin F$. By Proposition 3.21, $x^* \in \sigma(F)$. Conversely, let $\sigma(x) \notin F$. To prove that F is a σ-obstinate filter, we need only to show that $x^* \in \sigma(F)$. By hypothesis, we get that $x^* \in \sigma(F)$. Hence F is a σ-obstinate filter of L. □

The following example, shows that the condition of Theorem 3.26 is necessary.

Example 3.27. Consider Example 3.2(*ii*) when $\sigma(a) = 0$, $\sigma(b) = 1$ and $\sigma(c) = 0$. Consider $F = \{c, 1\}$. It is easy to verify that $\sigma(x) \in F$ or $x^* \in \sigma(F)$, for all $x \in L$ and $\sigma(F)$ isn't an upper set and F isn't σ-obstinate, since $\sigma(a) \notin F$, $\sigma(c) \notin F$ but $c \to a = b \notin \sigma(F)$.

Theorem 3.28. *Suppose that F and G are two proper filters of L such that $F \subseteq G$. If F is a σ-obstinate filter, then G is also σ-obstinate filter.*

Proof. Let $\sigma(x) \notin G$, $\sigma(y) \notin G$, for $x, y \in L$. So $\sigma(x), \sigma(y) \notin F$ then $x \to y \in \sigma(F)$, $y \to x \in \sigma(F)$, since F is a σ-obstinate filter. Hence $x \to y \in \sigma(G)$, $y \to x \in \sigma(G)$. Therefore G is a σ-obstinate filter of L. □

Using Theorem 3.28:

Corollary 3.29. *$\{1\}$ is a σ-obstinate filter of L if and only if every proper filter of L is a σ-obstinate filter.*

Corollary 3.30. *Let F be a proper σ-obstinate filter of L. Then $D(F)$ and $\mathrm{Rad}(F)$ are σ-obstinate filters of L.*

Proposition 3.31. *Let F, G and I be filters of L. If I is a σ-obstinate filter, $\sigma(I)$ is a proper filter and $F \cap G \subseteq \sigma(I)$, then $\sigma(F) \subseteq I$ or $\sigma(G) \subseteq I$.*

Proof. Let $F \cap G \subseteq \sigma(I)$, $\sigma(F) \nsubseteq I$ and $\sigma(G) \nsubseteq I$. We take $a \in \sigma(F) - I$ and $b \in \sigma(G) - I$, then

$$a \in \sigma(F), a \notin I \text{ and } b \in \sigma(G), b \notin I, \quad \textbf{(1)}.$$

So exist $a' \in F$, $b' \in G$ such that $a = \sigma(a')$ and $b = \sigma(b')$. As

$$a' \vee b' \in F \cap G \text{ and } F \cap G \subseteq \sigma(I), a' \vee b' \in \sigma(I), \quad \textbf{(2)}.$$

By applying Theorem 3.26, $a = \sigma(a') \in I$ or $(a')^* \in \sigma(I)$. Now by **(1)**, $a \notin I$, so we obtain $(a')^* \in \sigma(I)$ and $(b')^* \in \sigma(I)$. Since $\sigma(I)$ is a filter, we get $(a')^* \wedge (b')^* \in \sigma(I)$. By Theorem 2.3, we know that $(a'^* \wedge b'^*) = (a' \vee b')^*$. Hence $(a' \vee b')^* \in \sigma(I)$. Therefore $(a' \vee b') \to 0 = (a' \vee b')^* \in \sigma(I)$. By **(2)**, $a' \vee b' \in \sigma(I)$. Since $\sigma(I)$ is a filter, we get that $0 \in \sigma(I)$. That is a contradiction, therefore $\sigma(F) \subseteq I$ or $\sigma(G) \subseteq I$. $\qquad\square$

Theorem 3.32. *Let F be a σ-obstinate σ-filter of L such that $\sigma(F)$ is an upper set of L. Then every proper filter of quotient algebra L/F is a $\overline{\sigma}$-obstinate filter.*

Proof. Assume that F is a σ-obstinate filter of L and $x \in L$ such that $\overline{\sigma}([x]) \notin \{[1]\}$, i.e. $[\sigma(x)] \neq [1]$, then $\sigma(x) \notin F$. We apply that hypothesis, and obtain $x^* \in \sigma(F)$, then $x^* \in F$. So $[x^*] = [1] \in \overline{\sigma}([1])$. Based on Theorem 3.26, $\{[1]\}$ is a $\overline{\sigma}$-obstinate filter of L/F. Finally, by using Corollary 3.29, we get that every filter of a quotient algebra L/F is a $\overline{\sigma}$-obstinate filter. $\qquad\square$

Proposition 3.33. *Let F be a σ-obstinate σ-filter of L such that $\sigma(F)$ is an upper set and $\sigma(x^*) \notin F$, for all $0 \neq x \in L$. Then L/F is a special BL-algebra.*

Proof. Let $[0] \neq [x] \in L/F$. If $\sigma(x^*) \notin F$, for all $0 \neq x \in L$ and F is a σ-obstinate σ-filter, we can get that $x^{**} \in \sigma(F) \subseteq F$, then $[x^*] = [0]$, hence L/F is a special BL-algebra. $\qquad\square$

According to Proposition 3.33:

Corollary 3.34. *Let $\{1\}$ be a σ-obstinate filter of L and $\sigma(x^*) \neq 1$, for all $0 \neq x \in L$. Then L is a special BL-algebra.*

The following example shows that the converse of Proposition 3.33 isn't true in general.

Example 3.35. Consider Example 3.2(*iii*), when $\sigma(a) = c$, $\sigma(b) = 1$ and $\sigma(c) = c$. Consider $F = \{b, 1\}$, it is easy to verify that F is a σ-filter and L/F is a special BL-algebra, while F isn't a σ-obstinate filter, since $\sigma(a) \notin F$, $\sigma(0) \notin F$ but $a \to 0 = 0 \notin \sigma(F)$.

4 σ-integral BL-algebras and σ-integral filters

In this section, we introduce the notion of σ-integral filters in a BL-algebra. Also, we investigate some characterizations of these filters.

Definition 4.1. A BL-algebra L is called a σ-integral BL-algebra, if $\sigma(x) \odot \sigma(y) = 0$, then $x = 0$ or $y = 0$, for all $x, y \in L$.

Example 4.2. (i) Consider Example 3.2(ii), when $\sigma(a) = b$, $\sigma(b) = a$ and $\sigma(c) = c$. It is clear that L is a σ-integral BL-algebra.

(ii) Consider Example 3.2(i), when $\sigma(a) = b$ and $\sigma(b) = a$. L isn't a σ-integral, since $\sigma(a) \odot \sigma(b) = 0$ but $a \neq 0$, $b \neq 0$.

Theorem 4.3. *Let L be a Gödel algebra and F be a σ-strongly primary σ-filter of L. Then L/F is a $\overline{\sigma}$-integral BL-algebra.*

Proof. Let F be a σ-strongly primary filter of L and $\overline{\sigma}([x]) \odot \overline{\sigma}([y]) = [0]$, for $[x], [y] \in L/F$. Then according to Lemma 2.9, there exist $m, n \in \mathbb{N}$ such that $[x]^n = [0]$ or $[y]^n = [0]$ and $[x^n] = [0]$ or $[y^n] = [0]$. Now, since L is a Gödel algebra, then $x^n = x$, $y^n = y$ and so $[x] = [0]$ or $[y] = [0]$. $\square$

The following example shows that the converse of Theorem 4.3, isn't correct in general.

Example 4.4. Consider Gödel algebra L in Example 3.2(i). Also take σ-filter $F = \{a, 1\}$ where $\sigma(a) = 1$, $\sigma(b) = 0$. It is clear that L/F is a $\overline{\sigma}$-integral BL-algebra, while F isn't a σ-strongly primary filter of L, since $(\sigma(a) \odot \sigma(b))^* = 1 \in F$ but $(a^n)^* = b \notin \sigma(F)$ and $(b^n)^* \notin \sigma(F)$, for all $n \in \mathbb{N}$.

From Theorem 4.3, we get the following corollary:

Corollary 4.5. *Let L be a Gödel algebra and $\{1\}$ be a σ-strongly primary filter of L. Then L is a σ-integral BL-algebra.*

Lemma 4.6. *Every σ-integral BL-algebra is integral.*

Proof. Let L be a σ-integral BL-algebra and $x \odot y = 0$, for all $x, y \in L$. Then $\sigma(x \odot y) = 0$ and so $\sigma(x) \odot \sigma(y) = 0$. As L is a σ-integral BL-algebra, so $x = 0$ or $y = 0$. Therefore L is an integral BL-algebra. $\square$

The following example shows that the converse of Lemma 4.6, isn't correct in general.

Example 4.7. Consider Example 3.2(iii), when $\sigma(a) = 0$, $\sigma(b) = 1$ and $\sigma(c) = 0$. It is clear that L is an integral BL-algebra, while L isn't a σ-integral BL-algebra (since $\sigma(a) \odot \sigma(c) = 0$ but $a \neq 0$ and $c \neq 0$).

Definition 4.8. A proper filter F of L is called a σ-integral filter, if for all $x, y \in L$, $(\sigma(x) \odot \sigma(y))^* \in F$ implies $x^* \in \sigma(F)$ or $y^* \in \sigma(F)$.

Example 4.9. (*i*) Consider Example 3.2(*ii*), when $\sigma(a) = b$, $\sigma(b) = a$ and $\sigma(c) = c$. It is clear, $F = \{a, 1\}$ is a σ-integral filter of L.

(*ii*) Consider Example 3.2(*iv*), when $\sigma(a) = a$ and $\sigma(b) = 1$. $F = \{b, 1\}$ isn't a σ-integral filter, since $(\sigma(a) \odot \sigma(a))^* \in F$ but $a^* \notin \sigma(F)$.

Theorem 4.10. *Let F be a σ-integral σ-filter of L. Then L/F is a $\overline{\sigma}$-integral BL-algebra.*

Proof. Let F be a σ-integral filter and $\overline{\sigma}([x]) \odot \overline{\sigma}([y]) = [0]$, for $[x], [y] \in L/F$. Then $(\sigma(x) \odot \sigma(y))^* \in F$ and so $x^* \in \sigma(F) \subseteq F$ or $y^* \in \sigma(F) \subseteq F$. Hence $[x] = [0]$ or $[y] = [0]$. $\square$

The following example shows that the converse of Theorem 4.10, isn't corret in general.

Example 4.11. Consider Example 3.2(*i*), when $\sigma(a) = 1$ and $\sigma(b) = 0$. For filter $F = \{a, 1\}$, L/F is a $\overline{\sigma}$-integral BL-algebra, while F isn't a σ-integral filter of L (since $(\sigma(a) \odot \sigma(b))^* \in F$ but $a^* \notin \sigma(F)$ and $b^* \notin \sigma(F)$).

Based on Theorem 4.10:

Corollary 4.12. *Let $\{1\}$ be a σ-integral filter of L. Then L is a σ-integral BL-algebra.*

Theorem 4.13. *Let F and G be two proper σ-filters of L such that $F \subseteq G$ and F be a σ-integral filter. Then G is a σ-integral filter of L.*

Proof. Let $(\sigma(x) \odot \sigma(y))^* \in G$, for $x, y \in L$. As $((\sigma(x) \odot \sigma(y)) \odot (\sigma(x) \odot \sigma(y))^*)^* \in F$ and F is a σ-integral filter, we have $(x \odot y)^* \in \sigma(F)$ or $(x \odot y)^{**} \in \sigma(F)$. If $(x \odot y)^{**} \in \sigma(F)$, then $(x \odot y)^{**} \in \sigma(G)$. Hence $(\sigma(x) \odot \sigma(y))^{**} \in \sigma(\sigma(G)) \subseteq \sigma(G) \subseteq G$, since G is a σ-filter. It is contradiction with $(\sigma(x) \odot \sigma(y))^* \in G$. So $(x \odot y)^* \in \sigma(F)$. Hence $(\sigma(x) \odot \sigma(y))^* \in F$, since F is a σ-filter. As F is a σ-integral filter, we have $x^* \in \sigma(F)$ or $y^* \in \sigma(F)$. Therefore $x^* \in \sigma(G)$ or $y^* \in \sigma(G)$, i.e. G is a σ-integral filter of L. $\square$

According to Theorem 4.13:

Corollary 4.14. *$\{1\}$ is a σ-integral filter of L, if and only if any proper σ-filter of L is a σ-integral filter.*

Lemma 4.15. *Let F be a σ-integral σ-filter of L. Then F is an integral filter of L.*

Proof. Let $(x \odot y)^* \in F$, for all $x, y \in L$. So $(\sigma(x) \odot \sigma(y))^* \in \sigma(F) \subseteq F$. Then $x^* \in \sigma(F) \subseteq F$ or $y^* \in \sigma(F) \subseteq F$. Therefore F is an integral filter of L. $\square$

The following example shows that the converse of Lemma 4.15, isn't correct in general.

Example 4.16. Consider Example 3.2(i), when $\sigma(a) = 1$ and $\sigma(b) = 0$. Clearly, $F = \{a, 1\}$ is an integral filter, while F isn't a σ-integral filter (since $(\sigma(a) \odot \sigma(b))^* \in F$, but $a^* \notin \sigma(F)$ and $b^* \notin \sigma(F)$).

Theorem 4.17. *Every σ-integral filter is a σ-strongly primary filter of L.*

Proof. Take $n = 1$ in Definition 2.6(iii). Then the proof is clear. □

The following example shows that the converse of Theorem 4.17 isn't true in general.

Example 4.18. Consider Example 3.2(iv), when $\sigma(a) = a$, $\sigma(b) = 1$. It is clear that $F = \{b, 1\}$ is a σ-strongly primary, while isn't a σ-integral filter, since $(\sigma(a) \odot \sigma(a))^* \in F$ but $a^* \notin \sigma(F)$.

Definition 4.19. A BL-algebra L is called a σ-semi-integral, if $\sigma(x) \odot \sigma(y) = 0$, for $x, y \in L$, implies $x = 0$ or $y^n = 0$, for some $n \in \mathbb{N}$.

Example 4.20. (i) Consider Example 3.2(ii), when $\sigma(a) = b$, $\sigma(b) = a$ and $\sigma(c) = c$. Clearly, L is a σ-semi-integral BL-algebra.

(ii) Consider Example 3.2(i), when $\sigma(a) = 1$ and $\sigma(b) = 0$. L isn't a σ-semi-integral BL-algebra (since $\sigma(a) \odot \sigma(b) = 0$ but $a, a^n \neq 0$ and $b, b^n \neq 0$, for all $n \in \mathbb{N}$).

By using Definitions 4.19 and 4.1:

Lemma 4.21. *Any σ-integral BL-algebra is a σ-semi-integral BL-algebra.*

The converse of Lemma 4.21 does not hold in the general case. See the following example:

Example 4.22. Consider Example 3.2(iv), when $\sigma(a) = a$, $\sigma(b) = 1$. L is a σ-semi-integral BL-algebra while it isn't a σ-integral BL-algebra, since $\sigma(a) \odot \sigma(a) = 0$, but $a \neq 0$, $b \neq 0$.

Definition 4.23. A proper filter F of L is called a σ-semi-integral filter, if $(\sigma(x) \odot \sigma(y))^* \in F$, implies $x^* \in \sigma(F)$ or $(y^n)^* \in \sigma(F)$, for some $n \in \mathbb{N}$ and for any $x, y \in L$.

Example 4.24. (i) Consider Example 3.2(ii), when $\sigma(a) = b$, $\sigma(b) = a$ and $\sigma(c) = c$. It is clear that $F = \{b, 1\}$ is a σ-semi-integral filter of L.

(ii) Consider Example 3.2(i), when $\sigma(a) = 0$, $\sigma(b) = 1$ and $\sigma(c) = 1$. $F = \{a, c, 1\}$ isn't a σ-semi-integral filter, since $(\sigma(a) \odot \sigma(b))^* \in F$ but $a^* \notin \sigma(F)$, $(b^n)^* \notin \sigma(F)$, for all $n \in \mathbb{N}$.

Lemma 4.25. *Any σ-semi-integral BL-algebra is semi-integral.*

Proof. Let L be a σ-semi-integral BL-algebra and $x \odot y = 0$, for all $x, y \in L$. So $\sigma(x \odot y) = 0$, hence $\sigma(x) \odot \sigma(y) = 0$. As L is a σ-semi-integral BL-algebra, $x = 0$ or $y^n = 0$, for some $n \in \mathbb{N}$. Therefore L is a semi-integral BL-algebra. $\square$

The following example shows that the converse of Lemma 4.25, isn't correct in general.

Example 4.26. Consider Example 3.2(iv), when $\sigma(a) = 0$, $\sigma(b) = 1$ and $\sigma(c) = 0$. It is clear that L is a semi-integral BL-algebra, while L isn't a σ-semi-integral BL-algebra (since $\sigma(a) \odot \sigma(c) = 0$ but $a \neq 0$ and $c^n \neq 0$, for all $n \in \mathbb{N}$).

Lemma 4.27. *Every σ-semi-integral σ-filter in BL-algebras is semi-integral.*

Proof. Let F be a σ-semi-integral σ-filter of L and $(x \odot y)^* \in F$, for all $x, y \in L$. So $(\sigma(x) \odot \sigma(y))^* \in \sigma(F) \subseteq F$. Then $x^* \in \sigma(F) \subseteq F$ or $(y^n)^* \in \sigma(F) \subseteq F$, for some $n \in \mathbb{N}$. $\square$

The following example shows that the converse of a BL-algebra Lemma 4.27 isn't correct in general.

Example 4.28. Consider Example 3.2(i), when $\sigma(a) = 1$, $\sigma(b) = 0$. It is clearly, $F = \{a, 1\}$ is a semi-integral filter, while F isn't σ-semi-integral filter (since $(\sigma(a) \odot \sigma(b))^* \in F$ but $a^* \notin \sigma(F)$ and $(b^n)^* \notin \sigma(F)$, for all $n \in \mathbb{N}$).

Theorem 4.29. *Let F be a σ-semi-integral σ-filter of L. Then L/F is a $\overline{\sigma}$-semi-integral BL-algebra.*

Proof. Let F be a σ-semi-integral σ-filter and $\overline{\sigma}([x]) \odot \overline{\sigma}([y]) = [0]$, for $[x], [y] \in L/F$. Then $[\sigma(x) \odot \sigma(y)] = [0]$ and hence $(\sigma(x) \odot \sigma(y))^* \in F$. So $x^* \in \sigma(F) \subseteq F$ or $(y^n)^* \in \sigma(F) \subseteq F$, for some $n \in \mathbb{N}$. Hence $[x] = [0]$ or $[y^n] = [0]$, for some $n \in \mathbb{N}$, i.e. L/F is a $\overline{\sigma}$-semi-integral BL-algebra. $\square$

Theorem 4.30. *Let σ be a monomorphism of L. Then every linearly ordered BL-algebra is a σ-semi-integral BL-algebra.*

Proof. Let $\sigma(x) \odot \sigma(y) = 0$, for $x, y \in L$. As L is a linearly ordered BL-algebra, $\sigma(x) \leq \sigma(y)$ or $\sigma(y) \leq \sigma(x)$. Therefore $(\sigma(x))^2 \leq \sigma(x) \odot \sigma(y)$ or $(\sigma(y))^2 \leq \sigma(x) \odot \sigma(y)$ and so $(\sigma(x))^2 = 0$ or $(\sigma(y))^2 = 0$. We get that $\sigma(x^2) = 0$ or $\sigma(y^2) = 0$. Hence $x^2 = 0$ or $y^2 = 0$. Thus L is a σ-semi-integral BL-algebra. $\square$

The converse of Theorem 4.30, isn't true in general.

Example 4.31. Consider Example 3.2(ii), when $\sigma(a) = b$, $\sigma(b) = a$ and $\sigma(c) = c$. It is clear that σ is a monomorphism and L is a σ-semi-integral BL-algebra, while L isn't linearly ordered, since $a \not\leq b$ and $b \not\leq a$.

Proposition 4.32. *Let F and G be proper σ-filters of L such that $F \subseteq G$ and F be a σ-semi-integral filter. Then G is a σ-semi-integral filter of L.*

Proof. Let $(\sigma(x) \odot \sigma(y))^* \in G$, for $x, y \in L$. As $((\sigma(x) \odot \sigma(y)) \odot (\sigma(x)\sigma(y))^*)^* \in F$ and F is a σ-semi-integral filter, we get that $(x \odot y)^* \in \sigma(F)$ or $(((x \odot y)^*)^n)^* \in \sigma(F)$, for some $n \in \mathbb{N}$. If $(((x \odot y)^*)^n)^* \in \sigma(F)$, for some $n \in \mathbb{N}$, then $(((x \odot y)^*)^n)^* \in \sigma(G)$. Hence $(((\sigma(x) \odot \sigma(y))^*)^n)^* \in \sigma(\sigma(G)) \subseteq \sigma(G) \subseteq G$, since G is a σ-filter. This is in contradiction with $((\sigma(x) \odot \sigma(y))^*)^n \in G$. So $(x \odot y)^* \in \sigma(F)$ and then $(\sigma(x) \odot \sigma(y))^* \in F$, since F is a σ-filter. As F is a σ-semi-integral filter, we have $x^* \in \sigma(F)$ or $(y^n)^* \in \sigma(F)$, for some $n \in \mathbb{N}$. Therefore $x^* \in \sigma(G)$ or $(y^n)^* \in \sigma(G)$, for some $n \in \mathbb{N}$. Thus G is a σ-semi-integral filter of L. $\square$

By Proposition 4.32:

Corollary 4.33. $\{1\}$ *is a σ-semi-integral filter of L, if and only if any proper σ-filter of L, is a σ-semi-integral filter.*

By Theorem 4.29:

Corollary 4.34. *Let $\{1\}$ be a σ-semi-integral filter. Then L is a σ-semi-integral BL-algebra.*

Theorem 4.35. *Let F be a proper σ-filter of L. Then F is a σ-semi-integral filter if and only if every $\overline{\sigma}$-filter of L/F is a $\overline{\sigma}$-semi-integral filter.*

Proof. Let F be a σ-semi-integral and $(\overline{\sigma}([x]) \odot \overline{\sigma}([y]))^* \in \{[1]\}$. So $[(\sigma(x) \odot \sigma(y))^*] \in \{[1]\}$. Then $(\sigma(x) \odot \sigma(y))^* \in F$ and so $x^* \in \sigma(F)$ or $(y^n)^* \in \sigma(F)$, for some $n \in \mathbb{N}$. Thus $x^* \in F$ or $(y^n)^* \in F$, since F is a σ-filter of L. Hence $[x]^* \in \{[1]\}$ or $([y]^n)^* \in \{[1]\}$, for some $n \in \mathbb{N}$. Therefore $\{[1]\}$ is a $\overline{\sigma}$-semi-integral filter and by Corollary 4.33, every $\overline{\sigma}$-filter of the quotient algebra L/F is a $\overline{\sigma}$-semi-integral filter. The converse is clear, by Corollary 4.33. $\square$

By Definitions 4.1 and 4.19:

Lemma 4.36. *Every σ-semi-integral Gödel algebra is σ-integral.*

According to Definitions 4.23 and 4.8:

Lemma 4.37. *In any BL-algebra, every σ-integral filter is a σ-semi-integral filter.*

The following example shows that the converse of the Proposition 4.37 isn't true in general.

Example 4.38. Consider Example 3.2(iv), when $\sigma(a) = a$, $\sigma(b) = 1$. $F = \{b, 1\}$ is a σ-semi-integral, while isn't a σ-integral filter. Since $(\sigma(a) \odot \sigma(a))^* \in F$ but $a^* = a \notin \sigma(F)$.

Lemma 4.39. *Let F be a σ-semi-integral filter of L and $\sigma(F)$ be an upper set of L. Then F is σ-strongly primary.*

Proof. Let $(\sigma(x) \odot \sigma(y))^* \in F$, for $x, y \in L$. Then $x^* \in \sigma(F)$ or $(y^n)^* \in \sigma(F)$, for some $n \in \mathbb{N}$. As $x^n \leq x$, we have $(x^n)^* \in \sigma(F)$ or $(y^n)^* \in \sigma(F)$, for some $n \in \mathbb{N}$ and so F is a σ-strongly primary filter of L. $\square$

By the definitions of σ-integral filters, σ-strongly primary filters, σ-semi-integral filters and Gödel algebras, we get that:

Corollary 4.40. *Let F be a filter of a Gödel algebra L. The following conditions are equivalent:*
   (i) *F is a σ-integral filter of L;*
   (ii) *F is a σ-strongly primary filter of L;*
   (iii) *F is a σ-semi-integral filter of L.*

5 Conclusion

In this paper, we continue the idea of Najmi Dolat Abadi and Moghaderi [13] and we investigated the properties of σ-filters in BL-algebras. In addition, due to the great importance of filters in logical algebras, we introduced other types of σ-filters in these algebras and studied their properties. We have also proved or disproved the relationships between the types of σ-filters in these algebras with theorems or examples.

In our future work, we are going to define other important concepts on σ-filters in BL-algebras, with the aim of studying this algebraic structure as closely as possible. We hope this work would serve as a foundation for further studies on the structure of BL-algebras and develop corresponding many-valued logical systems.

Acknowledgement. The authors would like to thank the anonymous reviewers for their constructive suggestion and helpful comments, which enabled us to improve the presentation of our work.

References

[1] A. Borumand Saeid and S. Motamed, Some results in BL-algebras, Math. Logic Quartely, 55(6) (2009), 649-658.

[2] A. Borumand Saeid and S. Motamed, Normal filters in BL-algebras, World Applied Sci.J.7 (Special Issue Apple. math), (2009), 70-76.

[3] A. Borumand Saeid and S. Motamed, A new filter in BL-algebras, Journal of Intelligent and Fuzzy Systems, 27(6) (2014), 2949-2957.

[4] R. A. Borzooei and A. Paad, Integral filters and integral BL-algebras, Italian Journal of Pure and Applied Mathematics, 30 (2013), 303-316.

[5] D. Busneag and D. Piciu, BL-algebra of fractions relative to an ∧-closed system. In: Analeal Stiintificeale Universitatii Ovidius Constanta, Seria Mathematic, Vol XI, Fascicola, 1 (2003a), 39-48.

[6] D. Busneag and D. Piciu, On the lattice of deductive systems of a BL-algebra, Cent. Eurg. Math., 1(2) (2003b), 221-238.

[7] A. Dinola, Georgescu G and Iorgulescu A, Pseudo BL-algebra: part I. Mult Valued Log 8(5-6) (2002), 673-714.

[8] P. Hájek, Metamathematics of Fuzzy logic, Dordrecht: Kluwer Academic Publishers, (1998).

[9] M. Haveshki, A. Borumand Saeid and E. Eslami, Some type of filters in BL-algebra, Soft Computing, 10(8) (2006), 657-664.

[10] A. Iorgulescu, Algebras of logic as BCK-algebra, Bucharest University of Economics Bucharest, Romania, (2008).

[11] N. Mohtashamnia and A. Borumand Saeid, A special type of BL-algebras, Annals of University of Craiova, Mathematics and Computer Science Series Volume, 39(1) (2012), 8-20.

[12] S. Motamed, Semi-integral filters and semi-integral BL-algebras, Buletinul Academiei Cademiei Ademiei de Stiinte a Republicii Moldora. Mathematica, 1(86) (2018), 12-23.

[13] F. Najmi Dolat Abadi and J. Moghaderi, Filters by BL-homomorphisms, Soft Computing, 23 (2019), 9831-9841.

[14] Z. Parvizi, S. Motamed, F. Khaksar Haghani and J. Moghaderi, On σ-filters in BL-algebras, Submitted.

[15] E. Turunen and S. Sessa, Local BL-algebras, International j. Multiple-Valued logic, 6(1) (2001), 229-249.

[16] E. Turunen, Mathematics behind Fuzzy logic. physica Verlag, Heidelberg, (1999a).

[17] L. A. Zadeh, Fuzzy sets, Information and Control, 8 (1965), 338-353

[18] E. Turunen, S. Sessa, *Local BL-algebras*, Int. J. Mult-Valued logic, **6** (2001) 229-249.

       Received 12 July 2023

Polynomially defined strong d-ringoids

Akbar Rezaei*

Department of Mathematics, Payame Noor University, P.O.Box 19395-4697, Tehran, Iran
rezaei@pnu.ac.ir

Hee Sik Kim

Department of Mathematics, Hanyang University, Seoul, 04763, Korea
heekim@hanyang.ac.kr

Somayeh Borhani Nejad Rayeni

Department of Mathematics, Payame Noor University, P.O.Box 19395-4697, Tehran, Iran
borhani@pnu.ac.ir

Abstract

By considering the notion of K-nonvanishing on a field K, we define polynomially defined d-ringoids, and obtain some conditions for it to be a polynomially defined strong d-ringoid.

1 Introduction

In 1999, J. Neggers and H. S. Kim [3] introduced an algebra of logic as a d-algebra. They studied among relations between d-algebras and oriented digraphs. In 2009, P. J. Allen [1] constructed a multitude of d-algebras related to constructive function triples on the real numbers $\mathbb{R}$ and on integral domains D. In 2010, J. S. Han et al. [2] defined several special varieties of d-algebras, such as strong d-algebras, (weakly) selective d-algebras and pre-d-algebras. In 2022, J. Neggers et al. [4] introduced a new generalization of rings as ringoids, and discussed several properties of the left distributive ringoids over a field which are not rings and studied several various types of ringoids, and presented proper examples for (d-algebra, left zero, left distributive)

The authors are deeply grateful to the referee for the valuable suggestions.
*Sponsored by Payame Noor University

ringoids. In [5], A. Rezaei et al. introduced a notion of K-nonvanishing map on a field K, and obtained polynomially defined d-algebras, as well as some conditions for these to be strong d-algebras. Moreover, they investigated some relations between polynomially defined d-algebras with left-zero semigroups and β-algebras.

The motivation of this study consists algebraic and logical arguments. Polynomially defined strong d-ringoids introduced and investigated in the paper belong to a wide class of algebras of logic, they are an extension of well-known rings. This study is a continuation of the [4] and [5], were the property of ringoids and polynomial defined d-algebras were studied. Here we consider the notion of a K-nonvanishing on a field K, and define polynomially defined d-ringoids, and obtain some conditions for it to be a polynomially defined strong d-ringoid. Also, we prove that if $|K| \geq 2$, then polynomially defined d-ringoids are not left(right) distributive.

2 Preliminaries

A *d-algebra* ([3]) is a non-empty set X with a constant 0 and a binary operation "$\rightarrow$" satisfying the following axioms:

(I) $x \rightarrow x = 0$,

(II) $0 \rightarrow x = 0$,

(III) $x \rightarrow y = y \rightarrow x = 0$ implies $x = y$ for $x, y \in X$.

A *strong d-algebra* ([2]) is a non-empty set X with a constant 0 and a binary operation "$\rightarrow$" if it satisfies (I), (II) and the following axiom:

(IV) $x \rightarrow y = y \rightarrow x$ implies $x = y$

for $x, y \in X$.

An algebra $(X, \rightarrow, +, 0)$ of type $(2, 2, 0)$ is called a *ringoid* ([4]) if it satisfies the following conditions:

(i) $(X, +, 0)$ is an abelian group,

(ii) $(X, \rightarrow)$ is a groupoid.

A ringoid $(X, \rightarrow, +, 0)$ is called ([4])

- *left distributive* if $x \rightarrow (y + z) = (x \rightarrow y) + (x \rightarrow z)$,

- *right distributive* if $(x + y) \rightarrow z = (x \rightarrow z) + (y \rightarrow z)$,

616

- *distributive* if both the left and the right distributive laws hold

for $x, y, z \in X$.

A ringoid $(X, \rightarrow, +, 0)$ is called a *d-algebra ringoid* (briefly, *d-ringoid*) if $(X, \rightarrow, 0)$ is a d-algebra.

Example 2.1. ([4]) Let $(\mathbb{R}, +, \cdot, 0, 1)$ be the field of all real numbers. Define a binary operation "$\rightarrow$" on $\mathbb{R}$ by $x \rightarrow y := x(x - y)$ for $x, y \in \mathbb{R}$. Then $(\mathbb{R}, \rightarrow, +, 0)$ is a d-ringoid.

Let $(K, +, \cdot, 0, 1)$ be a field and let L be an extension field of K. A map $h : L \times L \rightarrow L$ is said to be *K-nonvanishing* ([5]) if $h(x, y) \neq 0$ for $x, y \in K$.

Example 2.2. ([5]) (i) Let K be the real field $\mathbb{R}$. Define a map $h : K \times K \rightarrow K$ by

$$h(x, y) := x^2 + y^2 + 1$$

for $x, y \in K$.
Then h is K-nonvanishing, since $h(x, y) = x^2 + y^2 + 1 > 0$ for $x, y \in K$.
Further, if we let $K := \mathbb{C}$ the complex numbers, then h is not $\mathbb{C}$-nonvanishing, since it is equal to 0 for $x = i$ and $y = 0$.

(ii) Let $K = \mathbb{Z}/\langle p \rangle$ be a field of integers modulo p, where p is prime and $p > 3$. If we define a map $h(x, y) := x^{p-1} + y^{p-1} + 1$ for $x, y \in K$, then $h(x, y) \neq 0$ for $x, y \in K$, since, for $x \in K$, we have

$$x^{p-1} = \begin{cases} 1 & \text{if } x \neq 0, \\ 0 & \text{otherwise.} \end{cases}$$

Then $x^{p-1} + y^{p-1} \in \{0, 1, 2\}$, and so $h(x, y) = x^{p-1} + y^{p-1} + 1 \in \{1, 2, 3\}$. Thus, $h(x, y) \neq 0$ for $x, y \in K$.

3 Main results

In this section, we discuss polynomially defined d-ringoids, and we obtain some conditions to be polynomially defined strong d-ringoids.

Theorem 3.1. *Let K be a field. Define a binary operation "$\rightarrow$" on K by*

$$x \rightarrow y := x^k (x - y)^l h(x, y)$$

for $x, y \in K$, where $k \geq 1$, $l \geq 1$ are integers and $h : K \times K \rightarrow K$ is K-nonvanishing. Then $(K, \rightarrow, 0)$ is a d-algebra.

617

Proof. (I) Given $x \in K$, we have $x \to x = x^k(x - x)^l h(x, x) = 0$.

(II) Given $x \in K$, we have $0 \to x = 0^k(0 - x)^l h(0, x) = 0$.

(III) Assuming $x \to y = y \to x = 0$ for some $x \neq y$ in K. Then $x^k(x - y)^l h(x, y) = 0$ and $y^k(y - x)^l h(y, x) = 0$. Since $x \neq y$ and $h(x, y) \neq 0 \neq h(y, x)$, we obtain $x^k = 0 = y^k$. It follows that $x = y = 0$, which is a contradiction. Hence $(K, \to, 0)$ is a d-algebra. $\qquad\square$

By applying Theorem 3.1, we obtain d-ringoids as follows:

Corollary 3.2. *Let K be a field. Define a binary operation "$\to$" on K by*

$$x \to y := x^k(x - y)^l h(x, y)$$

for $x, y \in K$, where $k \geq 1$, $l \geq 1$ are integers and $h : K \times K \to K$ is K-nonvanishing. Then $(K, \to, +, 0)$ is a d-ringoid.

It is necessary to find whether the d-ringoids discussed in Corollary 3.2 are left distributive. See the following theorems.

Theorem 3.3. *Let K be a field with $|K| = 2$. Define a binary operation "$\to$" on K by*

$$x \to y := x^k(x - y)^l h(x, y)$$

for $x, y \in K$, where $k \geq 1$, $l \geq 1$ are integers and $h : K \times K \to K$ is K-nonvanishing. Then $(K, \to, +, 0)$ is a d-ringoid which is right distributive, but not left distributive.

Proof. Using Corollary 3.2, we see that $(K, \to, +, 0)$ is a d-ringoid. Since K is a field of order 2, we get $K \cong \mathbb{Z}_2 = \{0, 1\}$. For right distributivity law, since $(K, +, 0)$ is an abelian group, we obtain $0 + 0 = 0$, $1 + 0 = 0 + 1 = 1$ and $1 + 1 = 0$. Also, since $char K = 2$, it follows that $h(1, 0) + h(1, 0) = 0$. We can see that:

$(0 + 0) \to 0 = 0 \to 0 = 0 = 0 + 0 = (0 \to 0) + (0 \to 0)$;

$(0 + 0) \to 1 = 0 \to 1 = 0 = 0 + 0 = (0 \to 1) + (0 \to 1)$;

$(0 + 1) \to 0 = 1 \to 0 = h(1, 0) = 0 + h(1, 0) = (0 \to 0) + (1 \to 0)$;

$(0 + 1) \to 1 = 1 \to 1 = 0 = 0 + 0 = (0 \to 1) + (1 \to 1)$;

$(1 + 1) \to 0 = 0 \to 0 = 0 = h(1, 0) + h(1, 0) = (1 \to 0) + (1 \to 0)$;

$(1 + 1) \to 1 = 0 \to 1 = 0 = 0 + 0 = (1 \to 1) + (1 \to 1)$.

Thus, $(K, \to, +, 0)$ is a right distributive d-ringoid.

Now, we observe that:

$$h(1, 0) = 1 \to 0 = 1 \to (0 + 0) \neq (1 \to 0) + (1 \to 0) = h(1, 0) + h(1, 0) = 0.$$

Hence it is not left distributive. $\qquad\square$

618

Theorem 3.4. *Let K be a field with $|K| \geq 2$. Define a binary operation "$\rightarrow$" on K by*

$$x \rightarrow y := x^k (x - y)^l h(x, y)$$

for $x, y \in K$, where $k \geq 1$, $l \geq 1$ are integers and $h : K \times K \rightarrow K$ is K-nonvanishing. Then $(K, \rightarrow, +, 0)$ is a d-ringoid which is not left distributive.

Proof. Assuming $(K, \rightarrow, +, 0)$ is a left distributive d-ringoid for some $k_1 \geq 1$ and $l_1 \geq 1$. Then we have

$$x \rightarrow (y + z) = (x \rightarrow y) + (x \rightarrow z).$$

for $x, y, z \in K$. It follows that

$$
\begin{aligned}
x^{k_1}(x - (y + z))^{l_1} h(x, y + z) &= x^{k_1}(x - y)^{l_1} h(x, y) + x^{k_1}(x - z)^{l_1} h(x, z) \\
&= x^{k_1}((x - y)^{l_1} h(x, y) + (x - z)^{l_1} h(x, z))
\end{aligned}
$$

for $x, y, x \in K$. If we let $y := x$ and $z := x$ for some $x \neq 0$ in K, then

$$
\begin{aligned}
x^{k_1}(x - (x + x))^{l_1} h(x, x + x) &= x^{k_1}((x - x)^{l_1} + (x - x)^{l_1}) h(x, x) \\
&= 0.
\end{aligned}
$$

This shows that

$$x^{k_1 + l_1} h(x, x + x) = 0$$

for some $x \neq 0 \in K$. Since h is K-nonvanishing, we get $x^{k_1 + l_1} = 0$. Since K is a field, we obtain $x = 0$, which leads to a contradiction. $\square$

Remark. We may give another proof of Theorem 3.4 as follows.
Assuming $(K, \rightarrow, +, 0)$ is a left distributive d-ringoid for some $k_1 \geq 1$ and $l_1 \geq 1$. Then we have

$$x \rightarrow (y + z) = (x \rightarrow y) + (x \rightarrow z).$$

for $x, y, z \in K$. If we let $x \neq 0$, $y := 0$ and $z := 0$ in K, then

$$
\begin{aligned}
x \rightarrow 0 &= x \rightarrow (0 + 0) \\
&= (x \rightarrow 0) + (x \rightarrow 0).
\end{aligned}
$$

It follows that $x \rightarrow 0 = 0$. On the other hand, we have

$$0 = x \rightarrow 0 = x^{k_1} x^{l_1} h(x, 0) = x^{k_1 + l_1} h(x, 0).$$

Since K is a field and h is K-nonvanishing, we get $x^{k_1 + l_1} = 0$, and hence $x = 0$, a contradiction.

Now, we construct d-ringoids which are not right distributive as below:

Theorem 3.5. *Let K be a field with $|K| > 2$. Define a binary operation "$\rightarrow$" on K by*

$$x \rightarrow y := x^k(x-y)^l h(x,y)$$

for $x, y \in K$, where $k \geq 1$, $l \geq 1$ are integers and $h : K \times K \rightarrow K$ is K-nonvanishing. Then $(K, \rightarrow, +, 0)$ is a d-ringoid which is not right distributive.

Proof. Similar to the proof of Theorem 3.4. $\qquad\qquad\qquad\qquad\qquad$ $\square$

It follows from Theorems 3.3, 3.4 and 3.5 that the following corollary holds.

Corollary 3.6. *Let K be a field with $|K| \geq 2$. Define a binary operation "$\rightarrow$" on K by*

$$x \rightarrow y := x^k(x-y)^l h(x,y)$$

for $x, y \in K$, where $k \geq 1$, $l \geq 1$ are integers and $h : K \times K \rightarrow K$ is K-nonvanishing. Then the d-ringoid $(K, \rightarrow, +, 0)$ is not distributive.

d-ringoids described in Theorem 3.1 are called *polynomially defined d-ringoids*. In addition, if d-algebra is strong, then it is called a *polynomially defined strong d-ringoid.*

Notice that if we change K-nonvanishing mapping h in Theorem 3.1, then we may obtain many different d-ringoids.

Example 3.7. Let $\mathbb{R}$ be the set of all real numbers. Define a binary operation "$\rightarrow$" on $\mathbb{R}$ by

$$x \rightarrow y := x^3(x-y)^2 h(x,y)$$

where $h : \mathbb{R} \times \mathbb{R} \rightarrow \mathbb{R}$ is $\mathbb{R}$-nonvanishing defined by $h(x,y) = x^2 + y^2 + 1$ for $x, y \in \mathbb{R}$. Then $(\mathbb{R}, \rightarrow, +, 0)$ is a polynomially defined strong d-ringoid.

It is known that every strong d-algebra is a d-algebra, but the converse need not be true in general (see [2]). By Theorem 3.1, we can find many polynomially defined d-ringoids. Among the polynomially defined d-ringoids, we want to find polynomially defined strong d-ringoids.

Theorem 3.8. *Let K be a field. Define a binary operation "$\rightarrow$" on K by*

$$x \rightarrow y := x^k(x-y)^l h(x,y)$$

for $x, y \in K$, where $k \geq 1$, $l \geq 1$ are integers and $h : K \times K \rightarrow K$ is K-nonvanishing. If the following conditions hold:

(i) k be an odd integer and l be an even integer,

(ii) $h(x, y) = h(y, x)$, *for* $x, y \in K$,

(iii) for $\alpha, \beta \in K$, $\alpha^k + \beta^k = 0$ *implies either* $\alpha = \beta$ *or* $\alpha = -\beta$,

then $(K, \to, +, 0)$ *is a polynomially defined strong d-ringoid.*

Proof. Assuming $x \to y = y \to x$. and $x \neq y$ then since l is an even integer $(x - y)^l = (-1)^l (y - x)^l = (y - x)^l$ and $h(x, y) = h(y, x)$, thus $x^k (x - y)^l h(x, y) = y^k (y - x)^l h(y, x)$, after cancellation by $(x - y)^l h(x, y)$, we get $x^k = y^k$, since k is an odd integer, $x^k + (-y)^k = 0$. By (iii) we obtain then $x = y$ or $x = -y$. If we assume $x = -y$, then $x^k + x^k = 2x^k = 0$, i.e., if $char K \neq 2$, this means $x^k = 0$, which implies $x = 0 = y$, if $char K = 2$, then $-y = y$. In both cases this leads to contradiction (with $x \neq y$). $\qquad \square$

Corollary 3.9. *Let K be a field. Define a binary operation "$\to$" on K by*

$$x \to y := x^k (x - y)^l h(x, y)$$

for $x, y \in X$, *where* $k \geq 1, l \geq 1$ *are integers and* $h : K \times K \to K$ *is K-nonvanishing. If the following conditions hold:*

(i) k be an even integer and l be an odd integer,

(ii) $h(x, y) = -h(y, x)$, *for* $x \neq y \in K$,

(iii) for $\alpha, \beta \in K$, $\alpha^k - \beta^k = 0$ *implies either* $\alpha = \beta$ *or* $\alpha = -\beta$,

then $(K, \to, +, 0)$ *is a strong d-ringoid.*

Proof. The proof is similar to Theorem 3.8, and we omit its proof. $\qquad \square$

Example 3.10. (i) Let $\mathbb{R}$ be the set of all real numbers. Define a binary operation "$\to$" on $\mathbb{R}$ by

$$x \to y := x^3 (x - y)^6 h(x, y)$$

where $h : \mathbb{R} \times \mathbb{R} \to \mathbb{R}$ is $\mathbb{R}$-nonvanishing defined by $h(x, y) = e^{x^2 + y^2}$ for $x, y \in \mathbb{R}$. Then $(\mathbb{R}, \to, +, 0)$ is a polynomially defined strong d-ringoid.

(ii) Consider the field $\mathbb{Z}_3$. Define a binary operation "$\to_1$" on $\mathbb{Z}_3$ by

$$x \to_1 y := x^5 (x - y)^4 h(x, y)$$

where $h : \mathbb{Z}_3 \times \mathbb{Z}_3 \to \mathbb{Z}_3$ is $\mathbb{Z}_3$-nonvanishing defined by $h(x, y) = (x + y)^2 + 1$ for $x, y \in \mathbb{R}$. Then $(\mathbb{Z}_3, \to_1, +, 0)$ is a polynomially defined strong d-ringoid.

(iii) Consider the field $\mathbb{Z}_3$. Define a binary operation "$\to_2$" on $\mathbb{Z}_3$ by

$$x \to_2 y := x^8 (x - y)^5 h(x, y)$$

where $h : \mathbb{Z}_3 \times \mathbb{Z}_3 \to \mathbb{Z}_3$ is $\mathbb{Z}_3$-nonvanishing defined by

$$h(x, y) = \begin{cases} x^3 - y^3 & \text{if } x \neq y, \\ 1 & \text{otherwise.} \end{cases}$$

for $x, y \in \mathbb{R}$. Then $(\mathbb{Z}_3, \to_2, +, 0)$ is a polynomially defined strong d-ringoid.

Theorem 3.11. *Let K be a field. Define a binary operation "$\to$" on K by*

$$x \to y := x^k (x - y)^l h(x, y)$$

for $x, y \in K$, where $k \geq 1$, $l \geq 1$ are integers and $h : K \times K \to K$ is K-nonvanishing. If the following conditions hold:

(i) l be an even integer,

(ii) $h(x, y) = h(y, x)$, for $x, y \in K$,

(iii) for $\alpha, \beta \in K$, $\alpha^k = \beta^k$ implies $\alpha = \beta$,

then $(K, \to, +, 0)$ is a polynomially defined strong d-ringoid.

Proof. It is enough to show (IV). Assuming $x \to y = y \to x$. Then since l is an even integer we have $(x - y)^l = (-1)^l (y - x)^l = (y - x)^l$ and $h(x, y) = h(y, x)$, thus $x^k (x - y)^l h(x, y) = y^k (y - x)^l h(y, x)$, after cancellation by $(x - y)^l h(x, y)$, we get $x^k = y^k$. By (iii) we obtain $x = y$. Hence $(K, \to, 0)$ is a strong d-algebra, and so $(K, \to, +, 0)$ is a polynomially defined strong d-ringoid. $\square$

Corollary 3.12. *Let K be a field. Define a binary operation "$\to$" on K by*

$$x \to y := x^k (x - y)^l h(x, y)$$

for $x, y \in K$, where $k \geq 1$, $l \geq 1$ are integers and $h : K \times K \to K$ is K-nonvanishing. If the following conditions hold:

(i) l be an odd integer,

(ii) $h(x, y) = -h(y, x)$, for $x \neq y \in K$,

(iii) for $\alpha, \beta \in K$, $\alpha^k = \beta^k$ implies $\alpha = \beta$,

then $(K, \rightarrow, +, 0)$ is a polynomially defined strong d-ringoid.

Proof. The proof is similar to Theorem 3.11, and we omit its proof. □

The following we pose an open problem:

Open Problem. How we can construct a field from a d-ringoid, specially, from a polynomially defined strong d-ringoid?

4 Conclusion

In this paper, we constructed polynomially defined d-ringoids on a field K. We found some suitable conditions for polynomially defined strong d-ringoids. Also, it is shown that for every field K with $|K| \geq 2$ the polynomially defined d-ringoids need not be left(right) distributive.

References

[1] P. J. Allen, *Constraction of many d-algebras*, Commun. Korean Math Soc., **24 (2)** (2009), 361–366.

[2] J. S. Han, H. S. Kim and J. Neggers, *Strong and ordinary d-algebras*, J. Multi-Valued Logic Soft Computing, **16** (2010), 331–339.

[3] J. Neggers and H. S. Kim, *On d-algebras,* Math. Slovaca, **49** (1999), 19–26.

[4] J. Neggers, H. S. Kim and A. Rezaei, *On ringoids,*, J. Algebr. Hyperstrucres Log. Algebr., **4** (2022), 37–50.

[5] A. Rezaei, and H. S. Kim, *Polynomially defined d-algebras and connections with β-algebras*, J. Multi-Valued Logic Soft Computing, to appear. In memory of Professor J. Neggers in [5]. We lost him, rather, his great name will be in our minds.

Received 8 July 2023

On Monadic Semi-Nelson Algebras

Shokoofeh Ghorbani

Department of Pure Mathematics, Faculty of Mathematics and Computer, Shahid Bahonar University of Kerman, Kerman, Iran
sh.ghorbani@uk.ac.ir

Abstract

In this paper, we equip the variety of semi-Nelson algebras with universal and existential quantifiers and introduce monadic semi-Nelson algebras. The monadic N-deductive systems and monadic congruences of monadic semi-Nelson algebras are defined and their properties are studied. We prove a one-to-one correspondence exists between the set of monadic congruences and the set of monadic N-deductive systems. The monadic semi-intuitionistic logic with strong negation is constructed. Based on the monadic semi-Nelson algebras, we prove the completeness and soundness of The monadic semi-intuitionistic logic with strong negation.

Keywords: Monadic semi-Nelson algebra, monadic N-deductive system, monadic semi-intuitionistic logic with strong negation, completeness.

1 Introduction

Nelson algebras, or N-lattices were introduced by Rasiowa [20] to provide an algebraic semantics to the constructive logic with strong negation proposed by Nelson in [16]. There is a close relationship between Nelson algebras and Heyting algebras. Vakarelov provided in [24] a way of constructing Nelson algebras from Heyting ones, by means of what is now known as a twist product.

Semi-Heyting algebra was introduced by Sankappanavar in [22] as a generalization of Heyting algebra. In [3], Cornejo and Viglizzo extended the Vakarelov's construction to semi-Heyting algebras obtained a new variety, which they called semi-Nelson algebras. The variety of semi-Nelson algebras is a natural generalization of Nelson algebras. In this variety, the lattice of congruences of an algebra is determined through some of its N-deductive systems. In [5], It was shown that there is an equivalence between the category of semi-Heyting algebras and the category of semi-Nelson algebras with center. (Also see[7])

Semi-intuitionistic logic with strong negation was introduced in [4] and proved to be complete with respect to the class of semi-Nelson algebras. An axiomatic extension is proved to have as algebraic semantics the class of Nelson algebras. (Also, see [2])

In 1955, Halmos introduced the concept of monadic Boolean algebras in [13]. He equipped Boolean algebras with a closure operator $\exists$ whose range is a subalgebra of Boolean algebra. The operator $\exists$ takes algebraic properties of the standard existential quantifier "for some." Since then, the monadic versions of the other algebraic structures have been studied by many researchers. Some important examples are monadic Heyting algebras [1], monadic MV-algebras [8], monadic BL-algebras [21], monadic residuated lattices[18] and their noncommutative cases [14], monadic bounded hoops [25] and Monadic pseudo equality-algebras [12].

This paper is organized as follows: In Section 2, some basic definitions and results are recalled. In Sect. 3, the notion of monadic semi-Nelson algebras is defined and some related properties are obtained. The monadic N-deductive systems and monadic congruences of monadic semi-Nelsons are defined and the relation between them is obtained and the monadic quotient semi-Nelson algebra is also defined. In Sect. 4, the monadic semi-intuitionistic logic with strong negation is presented and the soundness and completeness of this logic are proved based on monadic semi-Nelson algebras.

2   Preliminaries

In this section, we recall the basic definitions and some known results and properties about semi-Nelson algebras that we need in the rest of the paper.

Definition 2.1. ([3]) An algebra $\mathcal{A} = (A, \wedge, \vee, \rightarrow, \sim, 1)$ of type $(2, 2, 2, 1, 0)$ is called a semi-Nelson algebra if it satisfies the following conditions, for all $a, b, c \in A$:
(SN1) $a \wedge (a \vee b) = a$,
(SN2) $a \wedge (b \vee c) = (c \wedge a) \vee (b \wedge a)$,
(SN3) $\sim\sim a = a$,
(SN4) $\sim (a \wedge b) =\sim a\vee \sim b$,
(SN5) $a\wedge \sim a = (a\wedge \sim a) \vee (b\vee \sim b)$,
(SN6) $a \wedge (a \rightarrow_N b) = a \wedge (\sim a \vee b)$,
(SN7) $a \rightarrow_N (b \rightarrow_N c) = (a \wedge b) \rightarrow_N c$,
(SN8) $(a \rightarrow_N b) \rightarrow_N [(b \rightarrow_N a) \rightarrow_N [(a \rightarrow c) \rightarrow_N (b \rightarrow c)]] = 1$,
(SN9) $(a \rightarrow_N b) \rightarrow_N [(b \rightarrow_N a) \rightarrow_N [(c \rightarrow a) \rightarrow_N (c \rightarrow b)]] = 1$,
(SN10) $(\sim (a \rightarrow b)) \rightarrow_N (a\wedge \sim b) = 1$,
(SN11) $(a\wedge \sim b) \rightarrow_N (\sim (a \rightarrow b)) = 1$,

where $a \to_N b$ stands for the term $a \to (a \wedge b)$.

Proposition 2.2. ([3]) Let $\mathcal{A}$ be a semi-Nelson algebra. Then the following properties hold, for all $a, b, c \in A$:

(1) $a \wedge 1 = a$, $a \vee \sim 1 = a$,

(2) $a \to_N 1 = a \to_N a = 1$,

(3) $1 \to_N a = 1 \to a = a$,

(4) $\sim 1 \to_N a = 1$,

(5) $a \leq b$ if and only if $a \to_N b = 1$ and $\sim b \to_N \sim a = 1$,

(6) $a \leq b$ implies $c \to_N a \leq c \to_N b$ and $b \to_N c \leq a \to_N c$,

(7) $(a \to b) \to_N (a \to_N b) = 1$.

In the rest of the paper, ~ 1 is denoted by 0.

Definition 2.3. ([3]) Let $\mathcal{A}$ be a semi-Nelson algebra. A subset $D \subseteq A$ is called an N-deductive system of A if for all $a, b \in A$

(D1) $1 \in D$,

(D2) if $a, a \to_N b \in D$, then $b \in D$.

If D is an N-deductive system of A, then D is a lattice filter of A. For every subset X of A, the smallest deductive system of A containing X is called the N-deductive system generated by X and it will be denoted by $< X >$. The set $Ded_N(A)$ of all N-deductive systems of a semi-Nelson algebra $\mathcal{A}$, ordered by inclusion, form a complete lattice.

Proposition 2.4. ([3]) Let D be an N-deductive system of a semi-Nelson algebra $\mathcal{A}$. Then the binary relation defined on A by $x \equiv_D y$ if and only if

$$a \to_N b, b \to_N a, \sim a \to_N \sim b, \sim b \to_N \sim a \in D$$

is a congruence.

Theorem 2.5. ([3]) Let $\mathcal{A}$ be a semi-Nelson algebra. Then the lattices $Ded_N(A)$ and $Con(A)$ are isomorphic where $Con(A)$ is the congruence lattice of $\mathcal{A}$.

Proposition 2.6. ([3]) Let $\mathcal{A}$ be a semi-Nelson algebra and D be an N-deductive system of $\mathcal{A}$. Denote $A/D = \{[a]_D | a \in A\}$ where $[a]_D = \{b \in A | a \equiv_D b\}$. We define the following operations on A/D

$[a]_D \wedge [b]_D := [a \wedge b]_D$,

$[a]_D \vee [b]_D := [a \vee b]_D$,

$[a]_D \to [b]_D := [a \to b]_D$,

$\sim ([a]_D) = [\sim a]_D$.

Then $(A/D, \wedge, \vee, \to, \sim, [1]_D)$ is a semi-Nelson algebra.

3 Monadic semi-Nelson algebras

In this section, we will introduce monadic semi-Nelson algebras and obtain some of their properties.

Definition 3.1. An algebra $(A, \wedge, \vee, \rightarrow, \sim, \exists, \forall, 1)$ of type $(2, 2, 2, 1, 1, 1, 0)$ is called a monadic semi-Nelson algebra, if $(A, \wedge, \vee, \rightarrow, \sim, 1)$ is a semi-Nelson algebra and the following conditions are satisfied for all $a, b \in A$:

(MSN1) $\forall a \rightarrow_N a = 1$,
(MSN2) $a \rightarrow_N \exists a = 1$,
(MSN3) $\forall (a \rightarrow_N \exists b) = \exists a \rightarrow_N \exists b$,
(MSN4) $\forall (\exists a \rightarrow_N b) = \exists a \rightarrow_N \forall b$,
(MSN5) $\exists (a \wedge \exists b) = \exists a \wedge \exists b$,
(MSN6) $\forall (a \wedge b) = \forall a \wedge \forall b$,
(MSN7) $\exists \forall a = \forall a$,
(MSN8) $\exists \sim a = \sim \forall a$.

We denote a monadic semi-Nelson algebra $(A, \wedge, \vee, \rightarrow, \sim, \exists, \forall, 1)$ by $(A, \exists, \forall)$. The unary operators $\exists : A \rightarrow A$ and $\forall : A \rightarrow A$ are called an existential and a universal quantifiers on A, respectively.

Example 3.2. (1) It is easy to see that id_A is an existential (a universal) operator on any arbitrary semi-Nelson algebra $\mathcal{A}$. Thus every semi-Nelson algebra can be considered as a monadic semi-Nelson algebra.
(2) Let $A = \{0, 1, 2, 3, 4, 5, 6\}$ such that $4 < 3 < 5 < 6 < 2 < 0 < 1$. Consider the operations $\rightarrow$ and $\sim$ given by the following tables:

$\rightarrow$	0	1	2	3	4	5	6
0	1	0	2	3	4	5	6
1	0	1	2	3	4	5	6
2	0	0	1	5	5	5	6
3	0	0	0	1	1	1	1
4	0	0	0	1	1	1	1
5	0	0	0	1	1	1	1
6	0	0	0	1	1	1	1

$\sim$	0	1	2	3	4	5	6
	3	4	5	0	1	2	6

Then $\mathcal{A} = (A, \wedge, \vee, \rightarrow_N, \sim, 1)$ is a semi-Nelson algebra.([6])
Define $\forall$ and $\exists$ as follows:

$$\forall 0 = 0, \forall 1 = 1, \forall 2 = \forall 6 = 6, \forall 3 = \forall 5 = 3 \text{ and } \forall 4 = 4,$$
$$\exists 0 = \exists 2 = 0, \exists 1 = 1, \exists 3 = 3, \exists 4 = 4 \text{ and } \exists 5 = \exists 6 = 6.$$

It is easy to check that $\exists$ and $\forall$ are existential and universal quantifiers on A, respectively. Therefore $(A, \exists, \forall)$ is a monadic semi-Nelson algebra.

In the following proposition, we obtain some properties of existential and universal quantifiers on semi-Nelson algebras.

Proposition 3.3. Let $(A, \exists, \forall)$ be a monadic semi-Nelson algebra. Then the following hold for all $a, b \in A$:
(1) $\forall \sim a = \sim \exists a$,
(2) $\forall a \le a \le \exists a$,
(3) $\forall 1 = 1$, $\forall 0 = 0$,
(4) $\exists 1 = 1$, $\exists 0 = 0$,
(5) $\forall \exists a = \exists a$,
(6) $\forall \forall a = \forall a$,
(7) $\exists \exists a = \exists a$,
(8) if $a \le b$, then $\exists a \le \exists b$ and $\forall a \le \forall b$,
(9) $a \le \exists b$ if and only if $\exists a \le \exists b$,
(10) $\forall a \le b$ if and only if $\forall a \le \forall b$,
(11) $\exists a \le b$ if and only if $a \le \forall b$,
(12) $\forall a = a$ iff $\exists a = a$,
(13) $\forall a = 1$ iff $a = 1$,
(14) $\exists a = 0$ iff $a = 0$.

Proof. (1)The result follows from (MSN8) and (SN3).
(2) We have $\sim a \to_N \exists \sim a = 1$ by (MSN2). Thus $\sim a \to_N \sim \forall a = 1$ by (MSN8). Applying (MSN2) and Proposition 2.2 part (11), we get $\forall a \le a$. Also, using part (1) and then (MSN1), $\sim \exists a \to_N \sim a = \forall \sim a \to_N \sim a = 1$. By (MSN2) and Proposition 2.2 part (5), we obtain $a \le \exists a$.
(3) By (MSN2), (MSN2) and then Proposition 2.2 part (2), we have $\forall 1 = \forall (a \to_N \exists a) = \exists a \to_N \exists a = 1$. We have $\forall 0 \le 0$ by part (2). So $\forall 0 = 0$.
(4) Using (MSN8) and part (3), $\exists 0 = \exists \sim 1 = \sim \forall 1 = \sim 1 = 0$ and $\exists 1 = \exists \sim 0 = \sim \forall 0 = \sim 0 = 1$.
(5) By part (2), we have $\forall \exists a \le \exists a$. On the other hand, by (MSN8), part (1), (MSN4) and (MSN7), we obtain

$$\sim (\forall \exists a) \to_N \sim \exists a = \exists \forall \sim a \to_N \forall \sim a = \forall (\exists \forall \sim a \to_N \sim a)$$
$$= \forall (\forall \sim a \to_N \sim a) = \forall 1 = 1.$$

By (MSN4) and part (3), we have $\exists a \to_N \forall \exists a = \forall (\exists a \to_N \exists a) = \forall 1 = 1$. Thus $\exists a \le \forall \exists a$ by Proposition 2.2 part (5). Hence $\forall \exists a = \exists a$.

(6) Using (MSN7), (MSN4) and again part(MSN7), we obtain
$\forall a \rightarrow_N \forall\forall a = \exists\forall a \rightarrow_N \forall\forall a = \forall(\exists\forall a \rightarrow_N \forall a) = \forall(\forall a \rightarrow_N \forall a) = \forall 1 = 1.$
Applying twice (MSN8) and then (MSN2), we have

$$\sim \forall a \rightarrow_N \sim \forall\forall a = \exists \sim a \rightarrow_N \exists\exists \sim a = 1.$$

Thus $\forall a \leq \forall\forall a$ by Proposition 2.2 part (5). Also, we have $\forall\forall a \leq \forall a$ by part (2). Hence $\forall\forall a = \forall a$.

(7) By (MSN5) and part (3), we have $\exists\exists a = \exists(1 \wedge \exists a) = \exists 1 \wedge \exists a = \exists a$.

(8) By part (4) and assumption, we have $a \leq b \leq \exists b$. So $a = a \wedge \exists b$. Using (MSN5), $\exists a = \exists(a \wedge \exists b) = \exists a \wedge \exists b$. Hence $\exists a \leq \exists b$.
Since $\forall a \leq a \leq b$, then $\forall a = \forall\forall a = \forall(\forall a \wedge b) = \forall\forall a \wedge \forall b = \forall a \wedge \forall b$ part (6) and (MSN6). Hence $\forall a \leq \forall b$.

(9) Let $a \leq \exists b$. Then $\exists a \leq \exists\exists b = \exists b$ by part (2) and part (7). Conversely, let $\exists a \leq \exists b$. By part (2), we have $a \leq \exists a$. Hence $a \leq \exists b$.

(10) Suppose that $\forall a \leq b$. Then $\forall\forall a \leq \forall a$ by part (8). So $\forall a \leq \forall b$ by part (6). Conversely, let $\forall a \leq \forall b$. Since $\forall b \leq b$, then $\forall a \leq b$.

(11) Let $\exists a \leq b$. Then $\forall\exists a \leq \forall b$ by part (8). Applying part (5) and then part (2), we get $a \leq \forall b$. Conversely, let $a \leq \forall b$. Using part (8), (MSN7) and then part (2), we have $\exists a \leq \exists\forall b = \forall a \leq a$.

(12) Let $\forall a = a$. Then $\exists\forall a = \exists a$. By (MSN7), we get $\forall a = \exists a$. Since $\forall a = a$, then $a = \exists a$. Similarly, we can prove converse.

(13) Let $\forall a = 1$ for some $a \in A$. Since $\forall a \leq a$ by part (2), then $a = 1$. The converse follows from part (3).

(14) It follows from part (2) and part (4).

We recall that two isotone maps $f : P_1 \rightarrow P_2$ and $g : P_2 \rightarrow P_1$ form a residuated pair or an adjunction or a Galois connection between two poset P_1 and P_1 if $f(a) \leq b$ if and only if $a \leq g(b)$, for all $a \in P_1$ and $b \in P_2$.

Corollary 3.4. Let $(A, \exists, \forall)$ be a monadic semi-Nelson algebra. Then
(1) $(\exists, \forall)$ is a residuated pair over A,
(2) $\exists$ is a closure operator on A,
(3) $\forall$ is an interior operator on A,

Proof. (1) It follows from Proposition 3.3 part (11).
(2) The result follows from Proposition 3.3 parts (2), (7) and(8).
(3) It follows from Proposition 3.3 parts (2), (6) and(8).

Proposition 3.5. Let $(A, \exists, \forall)$ be a monadic semi-Nelson algebra. Then the following holds for all $a, b \in A$:

(1) $\exists(\exists a \wedge \exists b) = \exists a \wedge \exists b = \forall(\exists a \wedge \exists b),$
(2) $\exists(\forall a \wedge \forall b) = \forall a \wedge \forall b = \forall(\forall a \wedge \forall b),$
(3) $\forall(\exists a \vee b) = \exists a \vee \forall b,$
(4) $\forall(\forall a \vee b) = \forall a \vee \forall b,$
(5) $\forall(\exists a \vee \exists b) = \exists a \vee \exists b = \exists(\exists a \vee \exists b),$
(6) $\exists(a \vee b) = \exists a \vee \exists b.$

Proof. (1) Applying (MSN5) and Proposition 3.3 part (7), we get
$$\exists(\exists a \wedge \exists b) = \exists\exists a \wedge \exists b = \exists a \wedge \exists b.$$
By Proposition 3.3 part (12), we obtain $\forall(\exists a \wedge \exists b) = \exists a \wedge \exists b$.

(2) Using (MSN7) and (MSN5), we get
$$\exists(\forall a \wedge \forall b) = \exists(\forall a \wedge \exists\forall b) = \exists\forall a \wedge \exists\forall b = \forall a \wedge \forall b.$$
Proposition 3.3 part (12), we get $\forall(\forall a \wedge \forall b) = \forall a \wedge \forall b$.

(3) By (SN4), Proposition 3.3 part (1) and (3), (MSN5) and (MSN7),

$$\forall(\exists a \vee b) = \forall(\sim(\sim \exists a \wedge \sim b)) = \sim \exists(\sim \exists a \wedge \sim b)$$
$$= \sim \exists(\forall \sim a \wedge \sim b) = \sim \exists(\exists\forall \sim a \wedge \sim b)$$
$$= \sim(\exists\forall \sim a \wedge \exists \sim b) = \sim(\forall \sim a \wedge \exists \sim b)$$
$$= \sim(\sim \exists a \wedge \sim \forall b) = \exists a \vee \forall b$$

(4) We have $\forall(\forall a \vee b) = \forall(\exists\forall a \vee b) = \exists\forall a \vee \forall b = \forall a \vee \forall b$ by (MSN7) and part (3).

(5) We get $\forall(\exists a \vee \exists b) = \forall\exists a \vee \exists b = \exists a \vee \exists b$ by part (3) and then (MSN7). Using Proposition 3.3 part (12), we obtain $\exists(\exists a \vee \exists b) = \exists a \vee \exists b$.

(6) Since $a, b \leq a \vee b$, then $\exists a, \exists b \leq \exists(a \vee b)$ by Proposition 3.3 part (8). Hence $\exists a \vee \exists b \leq \exists(a \vee b)$. On the other hand, we have $a \leq \exists a$ and $b \leq \exists b$. So $a \vee b \leq \exists a \vee \exists b$. By part (5) and Proposition 3.3 part (8), we obtain $\exists(a \vee b) \leq \exists(\exists a \vee \exists b) = \exists a \vee \exists b$.

Proposition 3.6. Let $(A, \exists, \forall)$ be a monadic semi-Nelson algebra. Then the following statements hold for all $a, b \in A$:
(1) $\forall(a \rightarrow_N \forall b) = \exists a \rightarrow_N \forall b,$
(2) $\forall(\forall a \rightarrow_N b) = \forall a \rightarrow_N \forall b,$
(3) $\forall(\exists a \rightarrow_N \exists b) = \exists a \rightarrow_N \exists b = \exists(\exists a \rightarrow_N \exists b),$
(4) $\exists(\forall a \rightarrow_N \forall b) = \forall a \rightarrow_N \forall b = \forall(\forall a \rightarrow_N \forall b),$
(5) $\forall(\forall a \rightarrow_N \exists b) = \forall a \rightarrow_N \exists b = \exists(\forall a \rightarrow_N \exists b),$
(6) $\forall(a \rightarrow_N b) \leq \forall a \rightarrow_N \forall b,$
(7) $\forall(a \rightarrow_N b) \leq \exists a \rightarrow_N \exists b,$
(8) $\exists(a \rightarrow_N \exists b) \leq \forall(\forall a \rightarrow_N \exists b).$

Proof. (1) By (MSN7) and (MSN3), we have

$$\forall(a \to_N \forall b) = \forall(a \to_N \exists\forall b) = \exists a \to_N \exists\forall b = \exists a \to_N \forall b.$$

(2) Using parts (MSN7) and (MSN4), we get

$$\forall(\forall a \to_N b) = \forall(\exists\forall a \to_N b) = \exists\forall a \to_N \forall b = \forall a \to_N \forall b.$$

(3) We have $\forall(\exists a \to_N \exists b) = \exists\exists a \to_N \exists b = \exists a \to_N \exists b$ by (MSN3) and Proposition 3.3 part (7). So $\exists(\exists a \to_N \exists b) = \exists a \to_N \exists b$ by Proposition 3.3 part (12).

(4) Using (MSN7) and part (3), we obtain
$$\exists(\forall a \to_N \forall b) = \exists(\exists\forall a \to_N \exists\forall b) = \exists\forall a \to_N \exists\forall b = \forall a \to_N \forall b.$$
So $\forall(\forall a \to_N \forall b) = \forall a \to_N \forall b$ Proposition 3.3 part (12).

(5) We have $\forall(\forall a \to_N \exists b) = \exists\forall a \to_N \forall\exists b = \forall a \to_N \exists b$ by (MSN3) and (MSN7). Using Proposition 3.3 part (12), $\exists(\forall a \to_N \exists b) = \forall a \to_N \exists b$.

(6) We have $a \le \forall a$. By Proposition 2.2 part (6) $a \to_N b \le a \to_N \forall b$. Applying Proposition 3.3 part (8), (MSN7) and (MSN3), we get
$\forall(a \to_N b) \le \forall(a \to_N \forall b) = \forall(a \to_N \exists\forall b) = \exists\forall a \to_N \exists\forall b = \forall a \sim \forall b.$

(7) Since $b \le \exists b$, then $a \to_N b \le a \to_N \exists b$ By Proposition 2.2 part (6). Using Proposition 3.3 part (8) and then (MSN3), we have $\forall(a \to_N b) \le \forall(a \to_N \exists b) = \exists a \to_N \exists b$.

(8) We have $\forall a \le a$. Using Proposition 2.2 part (6) and (MSN3)
$$a \to_N \exists b \le \forall a \to_N \exists b = \exists\forall a \to_N \exists b = \forall(\forall a \to_N \exists b).$$
Using Proposition 3.3 part (8), we get $\exists(a \to_N \exists b) \le \exists\forall(\forall a \to_N \exists b) = \forall(\forall a \to_N \exists b)$.

Lemma 3.7. Let $(A, \exists, \forall)$ be a monadic semi-Nelson algebra. Then the following statements are equivalent:
(1) $a = \forall b$ for some b in A,
(2) $a = \exists b$ for some b in A,
(3) $a = \forall a$,
(4) $a = \exists a$,
(5) $\forall a = \exists a$.

Proof. Suppose that $a \in A$ is arbitrary.
$(1 \Rightarrow 2)$ Let $a = \forall b$ for some $b \in A$. Then $a = \exists\forall b$ by (MSN7).
$(2 \Rightarrow 3)$ If $a = \exists b$ for some $b \in A$, then $\forall a = \forall\exists b = \exists b = a$ by Proposition 3.3 part (5).
$(3 \Rightarrow 4)$ It follows from Proposition 3.3 part (12).
$(4 \Rightarrow 5)$ If $a = \exists a$, then $\forall a = \forall\exists a = \exists a = a$ by (MSN7).
$(5 \Rightarrow 6)$ Let $\forall a = \exists a$. Then $\forall a \le a \le \exists a$ by Proposition 3.3 part (2). Hence $a = \forall a$.

Corollary 3.8. Let $(A, \exists, \forall)$ be a monadic semi-Nelson algebra. Define $A_{\exists\forall} := \{a \in A | a = \exists a\}$ where $\exists A = \{\exists a | a \in A\}$ and $\forall A = \{\forall a | a \in A\}$. Then
(1) $A_{\exists\forall} = A$ iff $\forall = \exists = id_A$,
(2) $A_{\exists\forall} = \forall A_{\exists\forall} = \exists A_{\exists\forall} = \forall A = \exists A$.

Proof. It follows from Lemma 3.7.

Proposition 3.9. Let $(A, \exists, \forall)$ be a monadic semi-Nelson algebra. Then $A_{\exists\forall}$ is an N-subalgebra of $(A, \exists, \forall)$.

Proof. Let a, b be arbitrary elements of $A_{\exists\forall}$. Then $\exists a = a$ and $\exists b = b$. We have
(1) $\forall(a \wedge b) = \forall a \wedge \forall b = a \wedge b$ by (MSN6),
(2) $\forall(a \vee b) = \forall a \vee \forall b = a \vee b$ by Proposition 3.6 part (6),
(3) $\forall(a \rightarrow_N b) = \forall(\exists a \rightarrow_N \exists b) = \exists a \rightarrow_N \exists b = a \rightarrow_N b$ by Proposition 3.3 part (21),
(4) $\exists \sim a = \sim \forall a = \sim a$ by Proposition 3.3 part (10) and Lemma 3.7,
(5) $\exists \forall a = \forall a = a$ by Lemma 3.7,
(6) We have $\exists \exists a = \exists a = a$ by Proposition 3.3 part (9),
Thus $A_{\exists\forall}$ is closed under $\wedge$, $\vee$, $\rightarrow_N$, $\sim$, $\forall$ and $\exists$. Since $\exists 1 = 1$, then $1 \in A_{\exists\forall}$.

$A_{\exists\forall}$ is not necessarily a subalgebra of monadic semi-Nelson algebra in general since it may not be closed under $\rightarrow$. See the following example.

Example 3.10. Consider semi-Nelson algebra in Example 3.2 part (2) and define $\forall$ and $\exists$ as follows:

$$\forall 0 = \forall 2 = 2, \ \forall 1 = 1, \forall 3 = \forall 4 = 4, \ \forall 5 = 5 \text{ and } \forall 6 = 6,$$
$$\exists 0 = \exists 1 = 1, \ \exists 2 = 2, \ \exists 3 = \exists 5 = 5, \ \exists 4 = 4 \text{ and } \exists 6 = 6.$$

One can check that $\exists$ and $\forall$ are existential and universal quantifiers on A, respectively. Therefore $(A, \exists, \forall)$ is a monadic semi-Nelson algebra. Also, $A_{\exists\forall} = \{1, 2, 4, 5, 6\}$. Since $2 \rightarrow 1 = 0 \notin A_{\exists\forall}$, then $A_{\exists\forall}$ is not a subalgebra of $(A, \exists, \forall)$.

Proposition 3.11. Let $(A, \exists, \forall)$ be a monadic semi-Nelson algebra. Then $\forall a = \sup\{b \in A_{\exists\forall} | b \leq a\}$ and $\exists a = \inf\{b \in A_{\exists\forall} | a \leq b\}$.

Proof. We have $\exists \forall a = \forall a$. Thus $\forall a \in A_{\exists\forall}$. Also $\forall a \leq a$. Hence $\forall a \in \{b \in A_{\exists\forall} | b \leq a\}$. Therefore $\forall a \leq \sup\{b \in A_{\exists\forall} | b \leq a\}$.
On the other hand, let $b \in A_{\exists\forall}$ such that $b \leq a$. Then $b = \forall b \leq \forall a$. Thus $\sup\{b \in A_{\exists\forall} | b \leq a\} \leq \forall a$. The other part proves in the similar way.

Definition 3.12. Let $(A, \exists, \forall)$ be a monadic semi-Nelson algebra and D be an N-deductive system of $\mathcal{A}$. Then D is called a monadic N-deductive system of $(A, \exists, \forall)$ if $a \in D$ implies $\forall a \in D$.

The following example shows that the monadic N-deductive systems exist and any N-deductive system of a monadic semi-Nelson algebra may not be monadic in general.

Example 3.13. (1) $\{1\}$ is a monadic N-deductive system of an arbitrary monadic semi-Nelson algebra $(A, \exists, \forall)$.
(2) Consider a monadic semi-Nelson algebra $(A, \exists, \forall)$ in Example 3.2 part (2). Then $D_1 = \{1, 0\}$ is an N-deductive system of A but it is not monadic, because $\forall 1 = 2 \notin D_1$. Also, $D_2 = \{0, 1, 2\}$ is a monadic N-deductive system of $(A, \exists, \forall)$.

Let $(A, \exists, \forall)$ be a monadic semi-Nelson algebra. For any nonempty subset X of A, we denote by $< X >_m$ the monadic N-deductive system of $(A, \exists, \forall)$ generated by X, that is, $< X >_m$ is the smallest monadic N-deductive system of $(A, \exists, \forall)$ containing X. The set of all monadic N-deductive systems of $(A, \exists, \forall)$ is denoted by $MDed_N(A)$.

Proposition 3.14. Let X be a nonempty subset of a monadic semi-Nelson algebra. $(A, \exists, \forall)$. Then
(1) $< X >= \{a \in A : (x_1 \wedge ... \wedge x_n) \to_N a = 1, \exists x_1, ..., x_n \in X, n \in \mathcal{N}\}$
   $= \{a \in A : x_1 \to_N (... \to_N (x_n \to_N a)...) = 1, \exists x_1, ..., x_n \in X, n \in \mathcal{N}\}$,
(2) $< X >_m = < X \cup \forall X >$.

Proof.(1) The proof is straightforward, and we hence omit the details.
(2) First, We will prove that $< X \cup \forall X >$ is a monadic N-deductive system. Let $x \in < X \cup \forall X >$ be arbitrary. If $x = 1$, then $\forall x = \forall 1 = 1 \in < X \cup \forall X >$. If $x \neq 1$, then there exists $x_1, x_2, ..., x_n \in X \cup \forall X$ such that $(x_1 \wedge ... \wedge x_n) \to_N x = 1$ by part(1). Using Proposition 3.3 part (8), Proposition 2.2 part (6) and then, we obtain

$$1 = \forall 1 = \forall((x_1 \wedge ... \wedge x_n) \to_N x) \leq \forall(x_1 \wedge ... \wedge x_n) \to_N \forall x$$
$$= (\forall x_1 \wedge ... \wedge \forall x_n) \to_N \forall x.$$

Since $\forall x_1, ..., \forall x_n \in X \cup \forall X$, then $\forall x \in < X \cup \forall X >$ by part (1). Hence $< X \cup \forall X >$ is a monadic N-deductive system.
Now, suppose that $x \in \forall X$. Then there exits $y \in X$ such that $x = \forall y$. Thus $\forall x = \forall \forall y = \forall y \in < X >_m$. Therefore $X \cup \forall X \subseteq < X >_m$. Hence $< X \cup \forall X > \subseteq < X >_m$. Since $< X >_m \subseteq < X \cup \forall X >$, then $< X >_m = < X \cup \forall X >$.

Corollary 3.15. Let $(A, \exists, \forall)$ be a monadic semi-Nelson algebra. Then $MDed_N(A)$ forms an algebraic lattice whose compact elements are precisely the finitely generated monadic N-deductive systems.

Proof. By Proposition 3.14, it is easy to see that the mapping $X \mapsto < X >_m$ is an algebraic closure operator on the power set of A, and we have $< X >_m = \bigcup \{< Y >_m : Y$ is a finite subset of $X\}$ for every $X \subseteq A$. Hence $MDed_N(A)$ forms an algebraic lattice whose compact elements are precisely the finitely generated monadic N-deductive systems.

Proposition 3.16. Let $(A, \exists, \forall)$ be a monadic semi-Nelson algebra. Then D is a monadic N-deductive system of $(A, \exists, \forall)$ if and only if $D = < D \cap A_{\exists \forall} >$.

Proof. Let D be a monadic N-deductive system of $(A, \exists, \forall)$ and $x \in D$ be arbitrary. Then $\forall x \in D$. We have $\forall \forall x = \forall x$ by Proposition 3.3 part (6). So $\forall x \in D \cap A_{\exists \forall} \subseteq < D \cap A_{\exists \forall} >$. By Proposition 3.3 part (2), $\forall x \leq x$. Thus $x \in < D \cap A_{\exists \forall} >$. Hence $D \subseteq < D \cap A_{\exists \forall} >$. On the other hand, $D \cap A_{\exists \forall} \subseteq D$. So $< D \cap A_{\exists \forall} > \subseteq D$. Hence $D = < D \cap A_{\exists \forall} >$.

Conversely, let $D = < D \cap A_{\exists \forall} >$ and $x \in D$ be arbitrary. If $x = 1$, then $\forall x = \forall 1 = 1 \in D$. Suppose that $x \neq 1$. Then there exists $x_1, x_2, ..., x_n \in D \cap A_{\exists \forall}$ such that $(x_1 \wedge ... \wedge x_n) \to_N x = 1$ by Proposition 3.14. Then

$$1 = \forall 1 = \forall ((x_1 \wedge ... \wedge x_n) \to_N x) \leq \forall (x_1 \wedge ... \wedge x_n) \to_N \forall x$$
$$\leq (x_1 \wedge ... \wedge x_n) \to_N \forall x.$$

So $(x_1 \wedge ... \wedge x_n) \to_N \forall x \in D$. Since $x_1 \wedge ... \wedge x_n \in D$. Then $\forall x \in D$. Thus D is a monadic deductive system of $(A, \exists, \forall)$.

Definition 3.17. Let $(A, \exists, \forall)$ be a monadic semi-Nelson algebra and θ be a congruence on $\mathcal{A}$. Then θ is called a monadic congruence on $(A, \exists, \forall)$ if $(a, b) \in \theta$ implies $(\forall a, \forall b) \in \theta$ for all $a, b \in A$.

Theorem 3.18. Let $(A, \exists, \forall)$ be a monadic semi-Nelson algebra. Then there exists a one to one correspondence between its monadic N-deductive systems and their monadic congruences.

Proof. Suppose that D is a monadic N-deductive system. Then θ_D is a congruence by Proposition 2.4. Let $(a, b) \in \theta_D$ be arbitrary. So
$a \to_N b, b \to_N a \in D$ and $\sim a \to_N \sim b, \sim b \to_N \sim a \in D$.
Since D is a monadic N-deductive system, then

$\forall(a \to_N b), \forall(b \to_N a), \forall(\sim a \to_N \sim b), \forall(\sim b \to_N \sim a) \in D$.

By Proposition 3.6 part (6) and part (7), we get

$\forall a \to_N \forall b, \forall b \to_N \forall a, \exists \sim a \to_N, \exists \sim b, \exists \sim b \to_N \exists \sim a \in D$.

Using (MSN8), we obtain that $(\forall a, \forall b) \in \theta_D$. Hence θ_D is a monadic congruence on $(A, \exists, \forall)$.

Conversely, let θ be a monadic congruence on $(A, \exists, \forall)$. Then $D_\theta = \{x \in A : (x, 1) \in \theta\}$ is an N-deductive system. Let $a \in D_\theta$ be arbitrary. Then $(a, 1) \in \theta$. So $(\forall a, 1) = (\forall a, \forall 1) \in \theta$. Hence $\forall a \in D_\theta$. It can be easily shown that there exists a one to one correspondence between its monadic N-deductive systems and their monadic congruences.

Definition 3.19. A monadic semi-Nelson algebra $(A, \exists, \forall)$ is said to be subdirectly irreducible if it has the least nontrivial monadic congruence.

Let $(A, \exists, \forall)$ be a subdirectly irreducible monadic semi-Nelson algebra and θ be the least nontrivial monadic congruence. Then by Theorem 3.18, there exists a monadic N-deductive system D such that $\theta_D = \theta$, which means that D is the least monadic N-deductive system of A such that $D \neq \{1\}$. Thus, we can conclude that a monadic semi-Nelson algebra is subdirectly irreducible if among the nontrivial monadic N-deductive system of $(A, \exists, \forall)$, there exists the least one.

Proposition 3.20. Let $(A, \exists, \forall)$ be a monadic semi-Nelson algebra. Then $(A, \exists, \forall)$ is subdirectly irreducible if and only if there exists an element $1 \neq a \in A$ such that $a \in < x >_m$ for all $1 \neq x \in A$.

Proof. Let $(A, \exists, \forall)$ be subdirectly irreducible. Then $\cap\{D \in MDed_N(A) : D \neq \{1\}\} \neq \{1\}$. So $\cap\{< X >_m \in MDed_N(A) : x \neq 1\} \neq \{1\}$. Then there exists $a \in \cap\{< x >_m \in MDed_N(A) : x \neq 1\}$ such that $a \neq 1$.

Conversely, suppose that $D \in MDed_N(A)$ such that $D \neq \{1\}$. Then there exists $x \in D$ such that $x \neq 1$. By assumption $a \in < x >_m$. Thus $a \in D$. Hence $a \in \cap\{D \in MDed_N(A) : D \neq \{1\}\}$, that is $\cap\{D \in MDed_N(A) : D \neq \{1\}\} = \{1\}$. Therefore $(A, \exists, \forall)$ is a subdirectly irreducible monadic semi-Nelson algebra.

Lemma 3.21. Let θ be a monadic congruence on a monadic semi-Nelson algebra $(A, \exists, \forall)$. Then $(\exists a, \exists b) \in \theta$ for all $(a, b) \in \theta$.

Proof. Suppose that $(a, b) \in \theta$. Then $(\sim \forall \sim a, \sim \forall \sim b) \in \theta$. So we get $(\exists a, \exists b) \in \theta$ by (MSN8).

Proposition 3.22. Let D be a monadic N-deductive system of a monadic semi-Nelson algebra $(A, \exists, \forall)$. Define $\forall_D : A/D \to A/D$ by $\forall_D([a]_D) = [\forall a]_D$ and

$\exists_D : A/D \to A/D$ by $\exists_D([a]_D) = [\exists a]_D$. Then $(A/D, \forall_D, \exists_D)$ is a monadic semi-Nelson algebra and is called the quotient monadic semi-Nelson algebra.

Proof. The proof is straightforward.

4 Monadic semi-intuitionistic logic with strong negation

In this section, we introduce a propositional Monadic semi-intuitionistic logic with strong negation and obtain some its properties.

Definition 4.1. (i) The language of propositional monadic semi-intuitionistic logic with strong negation $\mathcal{SN}_M$ is the language of propositional semi-intuitionistic logic with strong negation $\mathcal{SN}$ expanded by the unary connectives $\exists$ and $\forall$.

(ii) Formulas of $\mathcal{SN}_M$ are defined in the following way:
each formula of $\mathcal{SN}$ is a formula of $\mathcal{SN}_M$, if φ is a formula, then $\exists\varphi$ and $\forall\varphi$ are formulas. We use the following definitions:

$\varphi \to_N \psi := \varphi \to (\varphi \wedge \psi)$,

$\varphi \Rightarrow \psi := (\varphi \to_N \psi) \wedge (\sim \psi \to_N \sim \varphi)$,

$\varphi \Leftrightarrow \psi := (\varphi \Rightarrow \psi) \wedge (\psi \Rightarrow \varphi)$.

The set of all formulas of $\mathcal{SN}_M$ is denoted by $\mathcal{F}_M$.

(iii) The logical axioms of $\mathcal{SN}_M$ consist of the logical axioms of $\mathcal{SN}$,

(A1) $(\varphi \to_N \psi) \to_N ((\psi \to_N \chi) \to_N (\varphi \to_N \chi))$,

(A2) $(\varphi \to_N \psi) \to_N ((\varphi \to_N \chi) \to_N (\varphi \to_N (\psi \wedge \chi)))$,

(A3) $(\varphi \wedge \psi) \to_N \varphi$,

(A4) $(\varphi \wedge \psi) \to_N \psi$,

(A5) $\varphi \to_N (\varphi \vee \psi)$,

(A6) $\psi \to_N (\varphi \vee \psi)$,

(A7) $\sim (\varphi \vee \psi) \to_N \sim \varphi$,

(A8) $\sim (\varphi \vee \psi) \to_N \sim \psi$,

(A9) $(\varphi \to_N \chi) \to_N ((\psi \to_N \chi) \to_N ((\varphi \vee \psi) \to_N \chi))$,

(A10) $(\sim \varphi \to_N \sim \psi) \to_N ((\sim \varphi \to_N \sim \chi) \to_N (\sim \varphi \to_N \sim (\psi \vee \chi)))$,

(A11) $\varphi \Rightarrow (\sim\sim \varphi)$,

(A12) $(\sim\sim \varphi) \Rightarrow \varphi$,

(A13) $(\varphi \to_N \psi) \to_N [(\psi \to_N \varphi) \to_N [(\varphi \to \chi) \to_N (\psi \to \chi)]]$,

(A14) $(\varphi \to_N \psi) \to_N [(\psi \to_N \varphi) \to_N [(\chi \to \varphi) \to_N (\chi \to \psi)]]$,

(A15) $[(\varphi \wedge \psi) \to_N \chi] \Rightarrow [\varphi \to_N (\psi \to_N \chi)]$,

(A16) $(\sim (\varphi \wedge \psi)) \Rightarrow (\sim \varphi \vee \sim \psi)$,

(A17) $(\sim \varphi \vee \sim \psi) \Rightarrow (\sim (\varphi \wedge \psi))$,

(A18) $(\varphi \wedge (\sim \varphi \vee \psi)) \Rightarrow (\varphi \wedge (\varphi \to_N \psi))$,

637

(A19) $(\varphi \to_N (\psi \to_N \chi)) \Rightarrow ((\varphi \wedge \psi) \to_N \chi)$,

(A20) $(\sim (\varphi \to \psi)) \to_N (\varphi \wedge \sim \psi)$,

(A21) $(\varphi \wedge \sim \psi) \to_N (\sim (\varphi \to \psi))$,

(A22) $[\sim (\varphi \wedge ((\chi \wedge \varphi) \vee (\psi \wedge \varphi)))] \to_N [\sim (\varphi \wedge (\psi \vee \chi))]$,

(A23) $\top$,

plus the following axioms :

(MSI1) $\forall \varphi \to_N \varphi$,

(MSI2) $\varphi \to_N \exists \varphi$,

(MSI3) $\forall (\varphi \to_N \exists \psi) \Leftrightarrow \exists \varphi \to_N \exists \psi$,

(MSI4) $\forall (\exists \varphi \to_N \psi) \Leftrightarrow \exists \varphi \to_N \forall \psi$,

(MSI5) $\exists (\varphi \wedge \exists \psi) \Leftrightarrow \exists \varphi \wedge \exists \psi$,

(MSI6) $\forall (\varphi \wedge \psi) \Leftrightarrow \forall \varphi \wedge \forall \psi$,

(MSI7) $\exists \forall \varphi \Leftrightarrow \forall \varphi$,

(MSI8) $\exists \sim \varphi \Leftrightarrow \sim \forall \varphi$.

(iv) The inference rules of $\mathcal{SN}_M$ are N-Modus Ponens (N-MP): from φ and $\varphi \to_N \psi$, we infer ψ, and (Ne) from φ, we infer $\forall \varphi$.

Definition 4.2. Let $(A, \exists, \forall)$ be a monadic semi-Nelson algebra. An A-evaluation of formulas is a mapping $e : \mathcal{F}_M \to A$, defined as follows: for all formulas $\varphi, \psi \in \mathcal{F}_M$

(1) $e(\varphi \wedge \psi) = e(\varphi) \wedge e(\psi)$,

(2) $e(\varphi \vee \psi) = e(\varphi) \vee e(\psi)$,

(3) $e(\varphi \to \psi) = e(\varphi) \to e(\psi)$,

(4) $e(\sim \varphi) = \sim e(\varphi)$,

(5) $e(\forall \varphi) = \forall e(\varphi)$,

(6) $e(\exists \varphi) = \exists e(\varphi)$.

(7) $e(\top) = 1$,

A formula φ is an A-tautology if $e(\varphi) = 1$ for each A-evaluation $e : \mathcal{F}_M \to A$. Formulas φ and ψ are semantically equivalent if $e(\varphi) = e(\psi)$.

Lemma 4.3. All axioms of the monadic semi-intuitionistic logic with strong negation are A-tautologies for all monadic semi-Nelson algebras $(A, \exists, \forall)$.

Proof. Suppose that $(A, \exists, \forall)$ is an arbitrary monadic semi-Nelson algebra and $e : \mathcal{F}_M \to A$ is an arbitrary A-evaluation. We will prove the axioms (MSI1) and (MSI4) are A-tautology. The proof of the other axioms are similar.

(MSI1) We have $e(\forall \varphi \to_N \varphi) = e(\forall \varphi \to (\forall \varphi \wedge \varphi)) = e(\forall \varphi) \to (e(\forall \varphi) \wedge e(\varphi)) = \forall e(\varphi) \to_N e(\varphi) = 1$ by (MSN1).

(MSI4) It is easy to prove that $\varphi \Leftrightarrow \psi$ if and only if $e(\varphi) = e(\psi)$. Hence $e(\forall (\exists \varphi \to_N \psi) \Leftrightarrow (\exists \varphi \to_N \forall \psi)) = 1$ by (MSN4).

Lemma 4.4. The inference rules of monadic semi-intuitionistic logic with strong negation are sound in the following sense: let $e : \mathcal{F}_M \to A$ be a truth evaluation where $(A, \exists, \forall)$ is a monadic semi-Nelson algebra:
(1) if $e(\psi) = 1$ and $e(\varphi \to_N \psi) = 1$, then $e(\varphi) = 1$,
(2) $e(\varphi) = 1$ implies $\forall e(\varphi) = 1$.

Proof.(1) Suppose that $e(\psi) = 1$ and $e(\varphi \to_N \psi) = 1$. Then $e(\varphi) = 1 \to_N e(\varphi) = e(\varphi) \to_N e(\psi) = e(\varphi \to_N \psi) = 1$.
(4) It follows from Proposition 3.3 part (3).

A proof in $\mathcal{SN}_M$ is a sequence $\varphi_1, \ldots, \varphi_n$ of formulas such that each φ_i either is an axiom of $\mathcal{SN}_M$ or follows from some preceding $\varphi_j, \varphi_k (j, k < i)$ by inference rules. A formula is provable (notation $\vdash \varphi$) if it is the least member a proof in $\mathcal{SN}_M$. By the following theorem, each provable formula in $\mathcal{SN}_M$ is A-tautology for all monadic semi-Nelson algebra $(A, \exists, \forall)$.

Theorem 4.5. (Soundness) The monadic semi-intuitionistic logic with strong negation is sound.

Proof. We should prove that if φ is provable in $\mathcal{SN}_M$, then φ is an A-tautology for each monadic semi-Nelson algebra $(A, \exists, \forall)$. So we have to show that all axioms and the inference rules of $\mathcal{SN}_M$ are A-tautology. Hence it follows from Lemma 4.3 and Lemma 4.4.

Suppose that Γ is a theory i.e.,a set of formulas in $\mathcal{SN}_M$. A proof in a theory Γ is a sequence $\varphi_1, \ldots, \varphi_n$ of formulas whose each member is either an axiom $\mathcal{SN}_M$ or a member of Γ (spacial axiom) or follows from some preceding members of the sequence using the inference rules. $\Gamma \vdash \varphi$ means that φ is provable in Γ, that is the last member of a proof in Γ.

Lemma 4.6. Let $\Gamma \cup \{\varphi, \psi\} \subseteq \mathcal{F}_M$ be formulas. Then $\mathcal{SN}_M$ proves the following :
(1) $\Gamma \vdash \varphi \Rightarrow \psi$ then $\Gamma \vdash \varphi \to_N \psi$ and $\Gamma \vdash \sim \psi \to_N \sim \varphi$,
(2) If $\Gamma \vdash \varphi \Rightarrow \psi$ and $\Gamma \vdash \varphi$, then $\Gamma \vdash \psi$,
(3) If $\Gamma \vdash \varphi$ and $\Gamma \vdash \psi$, then $\Gamma \vdash \varphi \wedge \psi$,
(4) $\Gamma \vdash \varphi \wedge \psi \Rightarrow \varphi$ and $\Gamma \vdash \varphi \wedge \psi \Rightarrow \psi$,
(4) $\Gamma \vdash \varphi \Rightarrow \varphi$,
(5) If $\Gamma \vdash \varphi \Rightarrow \psi$ and $\Gamma \vdash \psi \Rightarrow \chi$, then $\Gamma \vdash \varphi \Rightarrow \chi$,
(6) $\Gamma \cup \{\varphi \Rightarrow \psi\} \vdash (\varphi \vee \chi) \Rightarrow (\psi \vee \chi)$,
(7) $\Gamma \cup \{\varphi \Rightarrow \psi\} \vdash (\varphi \wedge \chi) \Rightarrow (\psi \wedge \chi)$,
(8) $\Gamma \cup \{\varphi \Rightarrow \psi\} \vdash \sim \psi \Rightarrow \sim \varphi$,

(9) $\Gamma \cup \{\varphi \Rightarrow \psi, \psi \Rightarrow \varphi, \phi \Rightarrow \chi, \chi \Rightarrow \phi\} \vdash (\varphi \to \phi) \Rightarrow (\psi \to \chi)$,

(10) $\Gamma \vdash (\varphi \to \psi) \to_N (\varphi \to_N \psi)$.

Proof. The proofs of these parts in semi-intuitionistic logic with strong negation can be found in [2] and [4], and their validity in $\mathcal{SN}_M$ is a direct consequence of Definition 4.1.

Corollary 4.7. Let $\varphi, \psi \in \mathcal{F}_M$ be formulas. $\mathcal{SN}_M$ proves the following :

(1) $\vdash \varphi \Leftrightarrow \varphi$,

(2) $\{\varphi \Leftrightarrow \psi\} \vdash \psi \Leftrightarrow \varphi$,

(3) $\{\varphi \Leftrightarrow \psi, \psi \Leftrightarrow \chi\} \vdash \varphi \Leftrightarrow \chi$,

(4) $\{\varphi \Leftrightarrow \psi\} \vdash (\varphi \wedge \chi) \Leftrightarrow (\psi \wedge \chi)$,

(4) $\{\varphi \Leftrightarrow \psi\} \vdash (\varphi \vee \chi) \Leftrightarrow (\psi \vee \chi)$,

(5) $\{\varphi \Leftrightarrow \psi\} \vdash (\varphi \to \chi) \Leftrightarrow (\psi \to \chi)$,

(6) $\{\varphi \Leftrightarrow \psi\} \vdash (\chi \to \varphi) \Leftrightarrow (\chi \to \psi)$,

(7) $\{\varphi \Leftrightarrow \psi\} \vdash \sim \varphi \Leftrightarrow \sim \psi$,

Proof. The results follow from Lemma 4.6.

Proposition 4.8. Let $\varphi, \psi \in \mathcal{F}_M$ be formulas. $\mathcal{SN}_M$ proves the following :

(1) $\{\varphi \to_N \psi, \varphi \Leftrightarrow \chi\} \vdash \chi \to_N \psi$,

(2) $\{\varphi \to_N \psi, \psi \Leftrightarrow \chi\} \vdash \varphi \to_N \chi$,

(3) $\vdash \forall \sim \varphi \Leftrightarrow \sim \exists \varphi$,

(4) $\vdash \forall \exists \varphi \Leftrightarrow \exists \varphi$,

(5) $\{\varphi \to_N \psi\} \vdash \forall \varphi \to_N \forall \psi$,

(6) $\{\varphi \to_N \psi\} \vdash \exists \psi \to_N \exists \varphi$,

(7) $\{\varphi \Leftrightarrow \psi\} \vdash \forall \varphi \Leftrightarrow \forall \psi$,

(8) $\{\varphi \Leftrightarrow \psi\} \vdash \exists \varphi \Leftrightarrow \exists \psi$.

Proof. (1) Put $\Gamma := \{\varphi \to_N \psi, \varphi \Leftrightarrow \chi\}$. Using Lemma 4.6 part (4), (N-MP) and part (1), we have $\Gamma \vdash \chi \to_N \varphi$. Using (A1) and twice (N-MP), we obtain $\Gamma \vdash \chi \to_N \psi$.

(2) The proof is similar to part (1).

(3) Applying (MSI8), $\vdash \sim \forall \sim \varphi \Leftrightarrow \exists \sim\sim \varphi$. By (A11) and (A12), we obtain $\vdash \sim \forall \sim \varphi \Leftrightarrow \exists \varphi$. Using Corollary 4.7 part (7), $\vdash \sim\sim \forall \sim \varphi \Leftrightarrow \sim \exists \varphi$. Again by (A11) and (A12), we get $\vdash \forall \sim \varphi \Leftrightarrow \sim \exists \varphi$.

(4) By (MSI7), we have $\vdash \exists \forall \sim \varphi \Leftrightarrow \forall \sim \varphi$. Using part (3), we obtain $\vdash \sim \forall \exists \varphi \Leftrightarrow \sim \exists \varphi$. By Corollary 4.7 part (7), $\vdash \sim\sim \forall \exists \varphi \Leftrightarrow \sim\sim \exists \varphi$. Applying (A11) and (A12), we get $\vdash \forall \exists \varphi \Leftrightarrow \exists \varphi$.

(5) Put $\Gamma := \{\varphi \to_N \psi\}$. Using (MSI1), Lemma 4.6 part (10) and (N-MP), we

obtain $\Gamma \vdash \forall \varphi \rightarrow_N \varphi$. Applying (A1), $\Gamma \vdash \varphi \rightarrow_N \psi$ and twice (N-MP), $\Gamma \vdash \forall \varphi \rightarrow_N \psi$. By (Ne), $\Gamma \vdash \forall(\forall \varphi \rightarrow_N \psi)$. By part (3), we have $\Gamma \vdash \forall(\exists \forall \varphi \rightarrow_N \psi)$. Using (MSI4) and (N-MP), $\Gamma \vdash \exists \forall \varphi \rightarrow_N \forall \psi$. Hence $\Gamma \vdash \forall \varphi \rightarrow_N \forall \psi$ by part (1) and part (3).

(6) Put $\Gamma := \{\varphi \rightarrow_N \psi\}$. Applying (MSI1), Lemma ref4.9 part (10) and (N-MP), we get $\vdash \psi \rightarrow_N \exists \psi$. Using $\Gamma \vdash \varphi \rightarrow_N \psi$, (A1) and twice (N-MP), We have $\Gamma \vdash \varphi \rightarrow_N \exists \psi$. So $\Gamma \vdash \forall(\varphi \rightarrow_N \exists \psi)$ by (Ne). Applying (MSI4) and (N-MP), we get $\Gamma \vdash \exists \varphi \rightarrow_N \exists \psi$.

(7) Put $\Gamma := \{\varphi \Leftrightarrow \psi\}$. Then $\Gamma \vdash \varphi \rightarrow_N \psi$, $\Gamma \vdash \psi \rightarrow_N \varphi$, $\Gamma \vdash \sim \varphi \rightarrow_N \sim \psi$ and $\Gamma \vdash \sim \psi \rightarrow_N \sim \varphi$ by Lemma 4.6 part (4) and (part (1). Applying part (5), we get $\Gamma \vdash \forall \varphi \rightarrow_N \forall \psi$ and $\Gamma \vdash \forall \psi \rightarrow_N \forall \varphi$.
Using part (6), we get $\Gamma \vdash \exists \sim \varphi \rightarrow_N \exists \sim \psi$ and $\Gamma \vdash \exists \sim \psi \rightarrow_N \exists \sim \varphi$. Applying (MSI8) part (1) and part (2), we obtain $\Gamma \vdash \sim \forall \varphi \rightarrow_N \sim \forall \psi$ and $\Gamma \vdash \sim \forall \psi \rightarrow_N \sim \forall \varphi$. Hence $\Gamma \vdash \forall \varphi \Leftrightarrow \forall \psi$ by Lemma 4.6 part (3).

(8) It follows from part (7), Corollary 4.7 part (7) and (MSI8).

The following is the standard Lindenbaum Tarski technique.

Corollary 4.9. Let Γ be a theory over monadic semi-intuitionistic logic with strong negation. Denote $F_\Gamma = \{[\varphi]_\Gamma : \varphi \in \mathcal{F}_M\}$ where $[\varphi]_\Gamma = \{\psi \in \mathcal{F}_M | \Gamma \vdash \varphi \Leftrightarrow \psi\}$. Define
$[\varphi]_\Gamma \wedge [\psi]_\Gamma = [\varphi \wedge \psi]_\Gamma$,
$[\varphi]_\Gamma \vee [\psi]_\Gamma = [\varphi \vee \psi]_\Gamma$,
$[\varphi]_\Gamma \rightarrow [\psi]_\Gamma = [\varphi \rightarrow \psi]_\Gamma$,
$\sim [\varphi]_\Gamma = [\sim \varphi]_\Gamma$,
$\forall [\varphi]_\Gamma = [\forall \varphi]_\Gamma$,
$\exists [\varphi]_\Gamma = [\exists \varphi]_\Gamma$,
$1 = [\top]_\Gamma$.
Then the algebra $(F_\Gamma, \wedge, \vee, \rightarrow, \sim, \exists, \forall, 1)$ is a monadic semi-Nelson algebra.

Proof. By Corollary4.7 and Proposition 4.8, $\Leftrightarrow$ is a congruence relation on $\mathcal{F}_M$. Hence the operations $\wedge, \vee, \rightarrow, \sim, \exists, \forall, 1$ are well defined on F_Γ. Thus $(F_\Gamma, \wedge, \rightarrow, \sim, \exists, \forall, 1)$ is an algebra of type $(2, 2, 2, 1, 1, 1, 0)$.
It is easy to prove that $(F_\Gamma, \wedge, \vee, \rightarrow, \sim, \exists, \forall, 1)$ is a monadic semi-Nelson algebra, and hence we omit the details.

Theorem 4.10. (Completeness) The monadic semi-intuitionistic logic with strong negation is complete, i.e. the following are equivalent:
(1) $\vdash \varphi$,
(2) φ is an A-tautology for every monadic semi-Nelson algebra $(A, \exists, \forall)$.

Proof. (1)$\Rightarrow$ (2) It follows from Theorem 4.5.

(2)$\Rightarrow$ (1) By Corollary 4.9, $(F_\Gamma, \wedge, \vee, \rightarrow, \sim, \exists, \forall, 1)$ is a monadic semi-Nelson algebra for every theory Γ in $\mathcal{SN}_M$. Let Γ to be the set of all axioms of $\mathcal{SN}_M$. Thus φ is an F_Γ-tautology by assumption. Consider the mapping e defined by $e(p) = [p]_\Gamma$ for all propositional variables p. Then e is an evaluation from $\mathcal{F}_M$ to a monadic semi-Nelson algebra $(F_\Gamma, \wedge, \vee, \rightarrow, \sim, \exists, \forall, 1)$. By Definition 4.2, we have $e(\varphi) = [1]_\Gamma$. So $[\varphi]_\Gamma = [1]_\Gamma$. So $\Gamma \vdash \varphi \Leftrightarrow \top$. Therefore $\vdash \varphi$.

5 Conclusion and future work

In this paper, we investigated the monadic semi-Nelson algebras, algebraic structures obtained by endowing the semi-Nelson algebras with universal and existential quantifiers. Some arithmetical properties of monadic semi-Nelson algebras were proven and then the relationship between the monadic N-deductive systems and the monadic congruences was established. Subdirectly irreducible monadic semi-Nelson algebras were characterized by monadic N-deductive systems. Then the monadic semi-intuitionistic logic was defined and it was shown that the Lindenbaum–Tarski algebra of this propositional calculus is a monadic semi-Nelson algebra. In future works, we will study the Belluce-semilattice associated with a monadic semi-Nelson algebras. Also, we will investigate and study the predicate semi-intuitionistic logic and the corresponding algebras (the polyadic semi-Nelson algebras and the cylindric semi-Nelson algebras).

References

[1] Bezhanishvili, G., 1998. Varieties of monadic Heyting algebras. Part I. Studia Logica, 61, pp.367-402.

[2] Cornejo, J.M. and Viglizzo, I., 2017. Proofs of some Propositions of the semi-Intuitionistic Logic with Strong Negation. arXiv preprint arXiv:1708.09448.

[3] Cornejo, J.M. and Viglizzo, I., 2018. Semi-Nelson algebras. Order, 35(1), pp.23-45.

[4] Cornejo, J.M. and Viglizzo, I., 2018. Semi-intuitionistic logic with strong negation. Studia Logica, 106, pp.281-293.

[5] Cornejo, J.M. and San Martin, H.J., 2018. A categorical equivalence between semi-Heyting algebras and centered semi-Nelson algebras. Logic Journal of the IGPL, 26(4), pp.408-428.

[6] Cornejo, J.M. and San Martin, H.J., 2020. Dually hemimorphic semi-Nelson algebras. Logic Journal of the IGPL, 28(3), pp.316-340.

[7] Cornejo, J.M., Gallardo, A. and Viglizzo, I., 2021. A categorial equivalence for semi-Nelson algebras. Soft Computing, 25(22), pp.13813-13821.

[8] Di Nola, A. and Grigolia, R., 2004. On monadic MV-algebras. Annals of Pure and Applied Logic, 128(1-3), pp.125-139.

[9] Drăgulici, D.D., 2001. Quantifiers on BL-algebras, An. Univ. Bucur., Mat. Inform, 50, pp.29-42.

[10] Drăgulici, D.D., 2010. Polyadic BL-Algebras. A Representation Theorem. Journal of Multiple-Valued Logic & Soft Computing, 16.

[11] Georgescu, G. and Leustean, I., 2000. A representation theorem for monadic Pavelka algebras. J. Univers. Comput. Sci., 6(1), pp.105-111.

[12] Ghorbani, S., 2019. Monadic pseudo-equality algebras. Soft Computing, 23(24), pp.12937-12950.

[13] Halmos, P.R., 1955. Algebraic logic, I. Monadic boolean algebras. Compositio Mathematica, 12, pp.217-249.

[14] Halmos, P.R., 1962. Algebraic Logic, Chelsea Publ. Co., New York, 453, p.454.

[15] Liu, L. and Zhang, X., 2023. The Belluce-semilattice associated with a monadic residuated lattice. Soft Computing, 27(11), pp.6983-6998.

[16] Nelson, D., 1949. Constructible falsity. Journal of Symbolic Logic, 14(1), pp.16-26.

[17] Odintsov, S.P., 2004. On the representation of N4-lattices. Studia Logica, 76(3), pp.385-405.

[18] Rachůnek, J. and Švrček, F., 2008. Monadic bounded commutative residuated l-monoids. Order, 25(2), pp.157-175.

[19] Rachůnek, J. and Šalounová, D., 2013. Monadic bounded residuated lattices. Order, 30(1), pp.195-210.

[20] Rasiowa, H., 1969. N-lattices and constructive logic with strong negation. Journal of Symbolic Logic, 34(1).

[21] Revaz, G., 2006. Monadic BL-algebras. Georgian Mathematical Journal, 13, pp.267-276.

[22] Sankappanavar, H.P., 2007, May. Semi-Heyting algebras: an abstraction from Heyting algebras. In Actas del IX Congreso Dr. A. Monteiro (pp. 33-66).

[23] Schwartz, D., 1980. Polyadic MV-Algebras. Mathematical Logic Quarterly, 26(36), pp.561-564.

[24] Vakarelov, D., 1977. Notes on N-lattices and constructive logic with strong negation. Studia Logica: An International Journal for Symbolic Logic, 36(1/2), pp.109-125.

[25] Wang, J., Xin, X. and He, P., 2018. Monadic bounded hoops. Soft computing, 22, pp.1749-1762.

[26] Wang, J.T., She, Y.H., He, P.F. and Ma, N.N., 2023. On categorical equivalence of weak monadic residuated distributive lattices and weak monadic c-differential residuated distributive lattices. Studia Logica, 111(3), pp.361-390.

Received 17 September 2023

Interval Ideals(Atoms) of Interval $BCI-$algebras

Ali Kordi

School of Mathematics, Statistic and Computer Sciences, Collage of Sciences, University of Tehran, P.O. Box 14155-6455, Tehran, Iran,
ali.kordi@gmail.com

Reza Ameri

School of Mathematics, Statistic and Computer Sciences, Collage of Sciences, University of Tehran, P.O. Box 14155-6455, Tehran, Iran,
rameri@ut.ac.ir

Abstract

The aim of this paper is the study of ideals of interval $BCI-$algebras, denoted by $IBCI-$algebras, as a generalization of $BCI-$algebras. In this regards, we introduce interval ideals of an $IBCI-$algebras and obtain properties of them. Also, we study atoms of interval $BCI-$algebras. In particular, we give some equivalent conditions to determine interval atoms of an interval $BCI-$algebra, based on its degenerated elements.

1 Introduction

Using intervals instead of exact values can also be perceived from the fact that the amount of imprecision can be codified through intervals in terms of its width. Since the operations in Ł∞ are continuous, the resulting interval operations are correct and optimal, which means that imprecision stored in input intervals are controlled by such operations. The limited capacity of machines to store just a finite set of finitely represented objects constraints the automatic computation of structures in which a machine representation of some objects exceeds such capacity. In the case of real numbers, although most programs provide highly accurate results, rounding errors build up during each step in the computation and may produce results that are not even meaningful. Among many algebraic structures, algebras of logic form important class of algebras. Examples of these are (resp. (resp. pseudo)$MV-$algebras, (resp.

pseudo)$BL-$algebras, (resp. pseudo)$BCK-$algebras, (resp. pseudo)$BCI-$algebras and etc., which are strongly connected to the logics. For example, $BCI-$algebras introduced by K. Isaki in 1966 [10] have connections with $BCI-$logic being the $BCI-$system in combinatoric logic which has application in the language of functional programming.

The notion of pseudo $BCI-$algebras has been introduced by W. A. Dudek and Y. B. Jun in [4], as an extension of $BCI-$algebras, and it was investigated by several authors(for more details see [5, 6, 8, 9]). As it is well known pseudo $BCI-$algebras are algebraic models for some extensions of a non-commutative version of $BCI-$logic. These algebras have also connections with other algebras of logic such as pseudo $BCK-$algebras, pseudo $BL-$algebras as well as pseudo $MV-$algebras.

From the logical point of view, various ideals corresponds to various sets of provable formulas. Recently, R. Santiago and et al. introduced and studied interval $BCI-$algebras, as a generalization of $BCI-$algebras(for more details see [13]). In particular, they proved that interval $BCI-$algebras, denoted by $IBCI-$algebras, which is different to pseudo $BCI-$algebras. As it is well known one of important notion ideals plays an important role in study of algebras, such as $BCI-$algebras and $BCK-$algebras. This paper is devoted to developing the notions of ideals and atoms for $IBCI-$algebras. In this paper we follow [13], to introduce and study interval ideals and interval atoms for an interval $BCI-$algebras. In section 3, we investigate the properties of interval ideals for an arbitrary $IBCI-$algebra and obtain some important results of them. Finally, in section 4, the notion of interval atoms for an interval $IBCI-$algebra are given. In particular some characterization for interval atoms in an interval $BCI-$algebra are obtained.

2 Preliminaries

In this section we present some definitions and properties of $BCI-$algebras, which we need to development our paper.

A $BCI-$algebra is an algebra $(X; \rightarrow, \top)$ such that is satisfied in the following conditions:

(1) $(y \rightarrow z) \rightarrow ((z \rightarrow x) \rightarrow (y \rightarrow x)) = \top$,

(2) $x \rightarrow ((x \rightarrow y) \rightarrow y) = \top$,

(3) $x \rightarrow x = \top$,

(4) if $x \rightarrow y = \top$ and $y \rightarrow x = \top$ imply $x = y$.

for all $x, y, z \in X$. We can define a partial ordering $" \leq "$ on X by $x \leq y$ if and only if $y \to x = \top$.

The following statements are true in any $BCI-$algebra X:

(1.1) $z \to (y \to x) = y \to (z \to x)$,

(1.2) $\top \to x = x$,

(1.3) $y \to x \leq (z \to y) \to (z \to x)$,

(1.4) $y \to x = \top$ implies $z \to y \leq z \to x$,

(1.5) $(y \to x) \to \top = (y \to \top) \to (x \to \top)$,

(1.6) $((y \to x) \to x) \to x = y \to x$.

Proposition 2.1. *([13]) Given a (resp. $BCK-$) $BCI-$algebra $(A, \to, \top)$ such that $(A, \preceq)$ is a meet semilattice verifying:*

$x \to (y \wedge z) = (x \to y) \wedge (x \to z)$, *for all* $x, y, z \in A$, $(\mathbb{A}, \to, \Rightarrow, [\top, \top])$, *and* $\mathbb{A} = \{[\underline{X}, \overline{X}] : \overline{X}, \underline{X} \in A, \underline{X} \preceq \overline{X}\}$. *For* $X, Y \in \mathbb{A}$, *define:*

1. $X \twoheadrightarrow Y = [\overline{X} \to \underline{Y}, \underline{X} \to \overline{Y}]$.
2. $X \Rightarrow Y = [(\underline{X} \to \underline{Y}) \wedge (\overline{X} \to \overline{Y}), \underline{X} \to \overline{Y}]$.

Then $\twoheadrightarrow$ *is the best representation of* $\to$ *and the algebra* $(\mathbb{A}, \twoheadrightarrow, \Rightarrow, [\top, \top])$ *satisfies:*

(IBCI1) $X \twoheadrightarrow (Y \twoheadrightarrow Z) = Y \twoheadrightarrow (X \twoheadrightarrow Z)$,

(IBCI2) $X \Rightarrow (Y \Rightarrow Z) = Y \Rightarrow (X \Rightarrow Z)$,

(IBCI3) $X \twoheadrightarrow Y \lesssim (Z \twoheadrightarrow X) \Rightarrow (Z \twoheadrightarrow Y)$

(IBCI4) $[\top, \top] \twoheadrightarrow X = X$,

(IBCI5) $X \ll Y \lesssim Z$ *implies* $X \ll Z$,

(IBCI6) $X \lesssim Y \ll Z$ *implies* $X \ll Z$,

(IBCI7) $X \lesssim Y$ *and* $Y \lesssim X$ *imply* $X = Y$,

where $X \ll Y$ *if and only if* $X \twoheadrightarrow Y = [\top, \top]$ *and* $X \lesssim Y$ *if and only if* $X \Rightarrow Y = [\top, \top]$. *However, when A has at least one element different from* $\top$, *then* $(\mathbb{A}, \twoheadrightarrow, [\top, \top])$ *is not a BCI algebra.*

Definition 2.2. ([13]) *Given a $BCI-$(resp. $BCK-$)algebra $(A, \to, \top)$ such that $(A, \preceq)$ is verifying:*

$x \to (y \wedge z) = (x \to y) \wedge (x \to z)$, *for any* $x, y, z \in A$, *where for all* $x, y \in A; x \wedge y := (x \to y) \to y$. *The algebra obtained by the method used in Proposition*

2.1. $(\mathbb{A}, \twoheadrightarrow, \Rightarrow, [\top, \top])$, is called interval $BCI-$(resp. $BCK-$) algebra, $IBCI-$(resp. $IBCK-$)algebra, for short.

Proposition 2.3. *([13]) Given an $IBCI-$(resp. $IBCK-$)algebra, $\mathbb{A}$, and $X, Y \in$ $\mathbb{A}$, the following properties are satisfied:*

$$\mathfrak{G} - 1 \quad X \twoheadrightarrow Y = [\overline{X} \to \underline{X}, \top],$$

$$\mathfrak{G} - 2 \quad X \twoheadrightarrow Y = [\top, \top],$$

if and only if $\overline{X} \preceq \underline{Y}$.

Given an $IBCI-$(resp. $IBCK-$)algebra, $\mathbb{A}$, then $X \in \mathbb{A}$ is called degenerates, if $X = [u, u]$ for some $u \in A$.

EXAMPLE 1. Suppose $(G, \cdot, e)$ is an Abelian group with e as the unite element. Define a binary operation $\to$ on A by putting $y \to x = xy^{-1}$. Then $(G, \cdot, e)$ is a $BCI-$algebra. Also, $(G, \preceq)$ is a meet semilattice, verifying $x \to (y \wedge z) = (x \to y) \wedge (x \to z)$, for any $x, y, z \in G$, where for all $x, y \in G$; $x \preceq y \implies yx^{-1} = e$. Now, define $\mathbb{G} = \{[x, y] : x, y \in G, x \preceq y\}$ and for all $X = [x_1, y_1], Y = [x_2, y_2] \in \mathbb{G}$:

$$X \twoheadrightarrow Y = [y_1 x_2^{-1}, y_2 x_1^{-1}] \quad , \quad X \Rightarrow Y = [y_2^{-1} x_2^{-1}, y_2 x_1^{-1}].$$

Then $(\mathbb{G}, \twoheadrightarrow, \Rightarrow, [e, e])$ is an $IBCI-$algebra.

EXAMPLE 2. Let $A = \mathbb{R}^2$ and define binary operation $\to$ and binary relation $\preceq$ on A by

$$(z, w) \to (x, y) = (x - z, y - w) \quad , \quad (x, y) \preceq (z, w) \Leftrightarrow (z, w) \to (x, y) = (0, 0),$$

for all $(x, y), (z, w) \in A$. Then $(A, \to, (0, 0))$ is a $BCI-$algebra. Define $\mathbb{A} = \{[x, y] : x, y \in A, x \preceq y\}$ and for all $X = [x_1, y_1], Y = [x_2, y_2] \in \mathbb{A}$:

$$X \twoheadrightarrow Y = [x_2 - y_1, y_2 - x_1] \quad , \quad X \Rightarrow Y = [y_2 - y_1, y_2 - x_1].$$

By some calculation it is concluded $(\mathbb{A}, \twoheadrightarrow, \Rightarrow, [(0, 0), (0, 0)])$ is an $IBCI-$algebra.

Proposition 2.4. *([13]) $X \twoheadrightarrow X = [\top, \top]$ if and only if X is degenerate.*

Definition 2.5. Let $\mathbb{A}$ be an $IBCI-$algebra, then the set of all degenerated elements of $\mathbb{A}$ is defined by

$$U_d(\mathbb{A}) = \{X \in \mathbb{A} | X = [u, u]\}.$$

Proposition 2.6. *([13]) For an $IBCI-$(resp. $IBCK-$)algebra, $\mathbb{A}$, for all $X, Y, Z \in \mathbb{A}$ and for all $K_d, V_d \in U_d(\mathbb{A})$, the following properties are satisfied:*

$(\mathcal{C}-1)$ $(Y \twoheadrightarrow Z) \lesssim ((Z \twoheadrightarrow K_d) \twoheadrightarrow (Y \twoheadrightarrow K_d)),$

$(\mathcal{C}-2)$ $K_d \twoheadrightarrow ((K_d \twoheadrightarrow V_d) \twoheadrightarrow V_d) = [\top, \top],$

$(\mathcal{C}-3)$ $X \twoheadrightarrow K_d = X \Rightarrow K_d,$

$(\mathcal{C}-4)$ $[\top, \top] \lesssim X$ *implies* $X = [\top, \top],$

$(\mathcal{C}-5)$ $X \lesssim Y$ *implies* $Y \twoheadrightarrow Z \lesssim X \twoheadrightarrow Z,$

$(\mathcal{C}-6)$ $X \lesssim Y$ *implies* $Z \twoheadrightarrow X \lesssim Z \twoheadrightarrow Y,$

$(\mathcal{C}-7)$ $X \lesssim Y$ *and* $Y \lesssim Z$ *implies* $X \lesssim Z,$

$(\mathcal{C}-8)$ $X \lesssim Y \twoheadrightarrow K_d$ *implies* $Y \lesssim X \twoheadrightarrow K_d,$

$(\mathcal{C}-9)$ $X \twoheadrightarrow Y \lesssim (K_d \twoheadrightarrow X) \twoheadrightarrow (K_d \twoheadrightarrow Y),$

$(\mathcal{C}-10)$ $((Y \twoheadrightarrow K_d) \twoheadrightarrow K_d) \twoheadrightarrow K_d = Y \twoheadrightarrow K_d,$

$(\mathcal{C}-11)$ $X \twoheadrightarrow Y \lesssim (Y \twoheadrightarrow X) \twoheadrightarrow [\top, \top],$

$(\mathcal{C}-12)$ $(X \twoheadrightarrow Y) \twoheadrightarrow [\top, \top] = (X \twoheadrightarrow [\top, \top]) \twoheadrightarrow (Y \twoheadrightarrow [\top, \top]),$

$(\mathcal{C}-13)$ *If* $X \ll Y$, *then* $X \Rightarrow Y = [\top, \top],$

$(\mathcal{C}-14)$ *If* $X \twoheadrightarrow Y = [\top, \top]$, *then* $X \lesssim Y.$

3   Interval Ideals in $IBCI-$algebras

In this section we generalize the notion of ideals in $BCI-$ algebras to intervalization ideals in $IBCI-$algebras, and we obtain some properties of them. Also, assume that $\mathbb{A}$ is an IBCI-algebra

Definition 3.1. A nonempty subset $\mathbb{I}$ of $\mathbb{A}$ is called an ideal of $\mathbb{A}$ if it is satisfied in the following conditions:

(I_1) $[\top, \top] \in \mathbb{I}$

(I_2) $X \twoheadrightarrow Y, X \Rightarrow Y \in \mathbb{I}$ and $X \in \mathbb{I}$ imply $Y \in \mathbb{I}.$

EXAMPLE 3. Consider a $BCI-$algebra $X = \{0, 1, a\}$ with the following Cayley table:

*	0	1	a
0	0	0	a
1	1	0	a
a	a	a	0

Then $(\mathbb{X}, \twoheadrightarrow, \Rightarrow, [0, 0])$ is an $IBCI-$algebra, where for all $U = [x_1, y_1], V = [x_2, y_2]$:

$$U \twoheadrightarrow V = [x_2 * y_1, y_2 * x_1] \quad , \quad U \Rightarrow V = [(y_2 * y_1) \wedge (x_2 * x_1), y_2 * x_1].$$

Therefore $\mathbb{I} = \{[0, 0], [0, 1], [1, 0], [1, 1]\}$ is an interval ideal of $\mathbb{X}$.

Proposition 3.2. *Let $\mathbb{I}$ be an interval ideal of $\mathbb{A}$. If $(X \ll Y, X \precsim Y)$ and $(\overline{X} \precsim \underline{Y}, X \in \mathbb{I})$, then $Y \in \mathbb{I}$.*

Proof. Straightforward. □

Theorem 3.3. *Let $\mathbb{I}$ be an interval ideal of $\mathbb{A}$. If for all $X, Y \in \mathbb{A}$, $X \ll Y$ and $X \precsim Y$ and $X \in \mathbb{I}$, then $Y \in \mathbb{I}$.*

Proof. It is an immediate consequence of definition of interval ideals. □

Let for any interval A of $\mathbb{A}$ we define that $\downarrow A := \{X \in \mathbb{A} | A \precsim X\}$, $A \downarrow := \{X \in \mathbb{A} | A \ll X\}$, $\uparrow A := \{X \in \mathbb{A} | X \precsim A\}$ and $A \uparrow := \{X \in \mathbb{A} | X \ll A\}$.

Theorem 3.4. *For any interval A of $\mathbb{A}$, $\downarrow A$ is an interval ideal of $\mathbb{A}$ if and only if for all $X, Y, Z \in \mathbb{A}$,*

$$Z \precsim Y \Rightarrow X \quad , \quad Z \ll Y \twoheadrightarrow X \quad , \quad Z \ll Y \quad imply \quad Z \precsim X.$$

Proof. Assume that $\downarrow A$ be an interval ideal of $\mathbb{A}$ and for all $X, Y, Z \in \mathbb{A}$, $Z \precsim Y \Rightarrow X$ and $Z \precsim Y$ implies that $Y \Rightarrow X \in \downarrow Z$ and $Y \in \downarrow Z$. Also, by $(\mathcal{C} - 14)$, we have $Y \twoheadrightarrow X \in \downarrow Z$. Since $\downarrow Z$ is an interval ideal, $X \in \downarrow Z$, it means that $Z \precsim X$. The converse, is obvious. □

Theorem 3.5. *Let A be an interval in $\mathbb{A}$. Then $A \downarrow$ is an interval ideal of $\mathbb{A}$ if and only if for all $X, Y, Z \in \mathbb{A}$:*

$$Z \precsim Y \Rightarrow X \quad , \quad Z \ll Y \twoheadrightarrow X \quad , \quad Z \ll Y \quad imply \quad Z \ll X,$$

where $Y \Rightarrow X$ is degenerated.

Proof. By our assumption, consider $Y \Rightarrow X = U = [u, u]$. Then $Z \precsim U$ implies that $Z \ll U$, by $(\mathcal{C}_d - 3)$, i.e., $Y \Rightarrow X \in Z \downarrow$. Also, $Z \ll Y \twoheadrightarrow X$ and $Z \ll Y$, then $Y \twoheadrightarrow X \in Z \downarrow$ and $Y \in Z \downarrow$. Since $Z \downarrow$ is an interval ideal of $\mathbb{A}$, we have $X \in Z \downarrow$. Therefore, $Z \ll X$, as desired. □

Proposition 3.6. *If $U_d = [u, u]$ is degenerated, then for any $X \in \mathbb{A}$, $U_d \twoheadrightarrow X = U_d \Rightarrow X$.*

Proof. We have

$$U_d \twoheadrightarrow X = [u \to \underline{X}, u \to \overline{X}] = [(u \to \underline{X}) \wedge (u \to \overline{X}), u \to \overline{X}] = U_d \Rightarrow X.$$

□

Theorem 3.7. *For any interval A of $\mathbb{A}$, $\uparrow A$ is an interval ideal of $\mathbb{A}$ if and only if for all $X, Y, Z \in \mathbb{A}$,*

$$Y \Rightarrow X \precsim Z \ , \ Y \twoheadrightarrow X \ll Z \ , \ Y \ll Z \ \text{imply} \ X \precsim Z.$$

Proof. Assume that $\uparrow A$ be an interval ideal of $\mathbb{A}$ and for all $X, Y, Z \in \mathbb{A}$, $Y \Rightarrow X \precsim Z$ implies that $Y \Rightarrow X \in \uparrow Z$. Also, by $(\mathcal{C} - 14)$, we have $Y \twoheadrightarrow X \in \downarrow Z$ and $Y \in \downarrow Z$. Since $\uparrow Z$ is an interval ideal, $X \in \uparrow Z$, it means that $X \ll Z$. The converse immediately follows from definition. $\qquad\square$

Theorem 3.8. *For any interval A of $\mathbb{A}$, $A \uparrow$ is an interval ideal of $\mathbb{A}$ if and only if for all $X, Y, Z \in \mathbb{A}$,*

$$Y \Rightarrow X \precsim Z \ , \ Y \twoheadrightarrow X \ll Z \ , \ Y \ll Z \ \text{imply} \ X \ll Z.$$

where $Y \Rightarrow X$ is degenerated.

Proof. By assumption, let $Y \Rightarrow X = U_d = [u, u]$. Then $U_d \precsim Z$ implies that $U_d \ll Z$, by Proposition 3.6, i.e., $Y \Rightarrow X \in Z \uparrow$. Also, $Y \twoheadrightarrow X \ll Z$ and $Y \ll Z$, then $Y \twoheadrightarrow X \in Z \uparrow$ and $Y \in Z \uparrow$. Since $Z \uparrow$ is an interval ideal of $\mathbb{A}$, we have $X \in Z \downarrow$. Therefore $X \ll Z$.
The converse is obvious. $\qquad\square$

Theorem 3.9. *Let $\mathbb{I}$ be an interval ideal of $\mathbb{A}$ and for all $X, Y, Z \in \mathbb{A}$, such that $X, Y \in \mathbb{I}$. Then the following statements are satisfied:*
(i) $X \precsim Y \twoheadrightarrow Z$ and $X \ll Y \Rightarrow Z$ then $Z \in \mathbb{I}$,
(ii) $X \ll Y \twoheadrightarrow Z$ and $X \precsim Y \Rightarrow Z$ then $Z \in \mathbb{I}$.

Proof. (i) Assume that $X \precsim Y \twoheadrightarrow Z$ and $X \ll Y \Rightarrow Z$ and $X \in \mathbb{I}$. Then $X \Rightarrow (Y \twoheadrightarrow Z) = [\top, \top] \in \mathbb{I}$ and $X \twoheadrightarrow (Y \Rightarrow Z) = [\top, \top] \in \mathbb{I}$ and $X \in \mathbb{I}$. Therefore, $(Y \Rightarrow Z) \in \mathbb{I}$ and $(Y \twoheadrightarrow Z) \in \mathbb{I}$. Since $Y \in \mathbb{I}$ and $\mathbb{I}$ is an interval ideal, we have $Z \in \mathbb{I}$.
(ii) The proof is similar to part (i). $\qquad\square$

An $IBCI$–subalgebra of $\mathbb{A}$ is a subset $\mathbb{S}$ of $\mathbb{A}$ such that for all $X, Y \in \mathbb{S}$ satisfies $Y \twoheadrightarrow X \in \mathbb{S}$ and $Y \Rightarrow X \in \mathbb{S}$.

Theorem 3.10. *Let $\mathbb{J}$ be an $IBCI$–subalgebra of an $IBCI$–algebra $\mathbb{A}$. Then $\mathbb{J}$ is an interval ideal of $\mathbb{A}$ if and only if for all $X \in \mathbb{J}$ and $Y \in \mathbb{A} - \mathbb{J}$ implies $X \Rightarrow Y \in \mathbb{A} - \mathbb{J}$ and $X \twoheadrightarrow Y \in \mathbb{A} - \mathbb{J}$.*

Proof. Suppose $X \Rightarrow Y \notin \mathbb{A} - \mathbb{J}$. Then $X \Rightarrow Y \in \mathbb{J}$. Since $X \in \mathbb{J}$ and $\mathbb{J}$ is an interval ideal of $\mathbb{A}$, one has $Y \in \mathbb{J}$, which is a contradiction. Also, if $X \twoheadrightarrow Y \notin \mathbb{A} - \mathbb{J}$,

651

then $X \twoheadrightarrow Y \in \mathbb{J}$. Since $X \in \mathbb{J}$ and $\mathbb{J}$ is an interval ideal of $\mathbb{A}$, it is concluded $Y \in \mathbb{J}$. , a contradiction. Thus it follows that $X \Rightarrow Y \in \mathbb{A} - \mathbb{J}$ and $X \twoheadrightarrow Y \in \mathbb{A} - \mathbb{J}$. Conversely, Since $\mathbb{J}$ is an $IBCI-$subalgebra, we affirm that $[\top, \top] \in \mathbb{J}$. Let $X \Rightarrow Y \in \mathbb{J}$ and $X \twoheadrightarrow Y \in \mathbb{J}$ and $X \in \mathbb{J}$. If $Y \notin \mathbb{J}$, then by hypothesis, one has $X \Rightarrow Y \in \mathbb{A} - \mathbb{J}$ and $X \twoheadrightarrow Y \in \mathbb{A} - \mathbb{J}$, a contradiction. Therefore, $\mathbb{J}$ is an interval ideal of $\mathbb{A}$. $\qquad\square$

Letting $\xi(\mathbb{A}) = \{X \in \mathbb{A} | X \ll [\top, \top]\}$, $\xi_{U_d}(\mathbb{A}) = \{X \in \mathbb{A} | X \ll U_d\}$, and $\xi_d(\mathbb{A}) = \{U_d \in \mathbb{A} | U_d \ll [\top, \top]\}$. A $\xi(\mathbb{A})$ is said to be an $IBCK-$part of $\mathbb{A}$.

Theorem 3.11. $\xi(\mathbb{A})$ and $\xi_d(\mathbb{A})$ are $IBCI-$subalgebras of $\mathbb{A}$.

Proof. Assume that $X, Y \in \xi(\mathbb{A})$, then $X \ll [\top, \top]$ and $Y \ll [\top, \top]$. Therefore, by $(\mathcal{C} - 12)$, we have:

$$(X \twoheadrightarrow Y) \twoheadrightarrow [\top, \top] = (X \twoheadrightarrow [\top, \top]) \twoheadrightarrow (Y \twoheadrightarrow [\top, \top]) = [\top, \top],$$

then, $X \twoheadrightarrow Y \ll [\top, \top]$, i.e., $X \twoheadrightarrow Y \in \xi(\mathbb{A})$. Also, by the same method we can show that $\xi_d(\mathbb{A})$ is an IBCI subalgebra. $\qquad\square$

Theorem 3.12. *For all* $X, Y \in \zeta(\mathbb{A})$, $Y \twoheadrightarrow X \in \zeta(\mathbb{A})$.

Proof. Assume that $X, Y \in \zeta(\mathbb{A})$. Then $X \ll [\top, \top], Y \ll [\top, \top]$. Now, by $(\mathcal{C} - 3)$, $X \lesssim [\top, \top], Y \lesssim [\top, \top]$ and by $(\mathcal{C} - 6)$, we have $Y \twoheadrightarrow X \lesssim Y \twoheadrightarrow [\top, \top] = [\top, \top]$ and by $(\mathcal{C} - 3)$, $Y \twoheadrightarrow X \ll [\top, \top]$, i.e, $Y \twoheadrightarrow X \in \zeta(\mathbb{A})$. $\qquad\square$

Theorem 3.13. *The following statements are hold:*
(i) Let $X \in \xi_d(\mathbb{A})$ *and* $Y \in \mathbb{A} - \xi(\mathbb{A})$. *Then* $Y \twoheadrightarrow X, Y \Rightarrow X \in \mathbb{A} - \xi(\mathbb{A})$.
(ii) Let $X \in \xi_d(\mathbb{A})$ *and* $Y \in U_d(\mathbb{A}) - \xi_d(\mathbb{A})$. *Then* $Y \twoheadrightarrow X, Y \Rightarrow X \in U(\mathbb{A}) - \xi_d(\mathbb{A})$.

Proof. (*i*) Assume that $Y \twoheadrightarrow X \in \xi(\mathbb{A})$. Since $\xi(\mathbb{A})$ is subalgebra, we affirm that $X \twoheadrightarrow (Y \twoheadrightarrow X) \in \xi(\mathbb{A})$. Thus, we have:

$$
\begin{aligned}
[\top, \top] &= (X \twoheadrightarrow (Y \twoheadrightarrow X)) \twoheadrightarrow [\top, \top] \\
&= (Y \twoheadrightarrow (X \twoheadrightarrow X)) \twoheadrightarrow [\top, \top] \\
&= (Y \twoheadrightarrow [\top, \top]) \twoheadrightarrow [\top, \top],
\end{aligned}
$$

therefore,

$$
\begin{aligned}
[\top, \top] &= [\top, \top] \twoheadrightarrow [\top, \top] \\
&= ((Y \twoheadrightarrow [\top, \top]) \twoheadrightarrow [\top, \top]) \twoheadrightarrow [\top, \top] \\
&= Y \twoheadrightarrow [\top, \top],
\end{aligned}
$$

i.e., $Y \in \xi(\mathbb{A})$. This is a contradiction. Also, if $Y \Rightarrow X \in \xi(\mathbb{A})$, since $\xi(\mathbb{A})$ is subalgebra, we can obtain $X \Rightarrow (Y \Rightarrow X) \in \xi(\mathbb{A})$. Thus, we have:

$$
\begin{aligned}
[\top, \top] &= (X \Rightarrow (Y \Rightarrow X)) \Rightarrow [\top, \top] \\
&= (Y \Rightarrow (X \Rightarrow X)) \Rightarrow [\top, \top] \\
&= (Y \Rightarrow [\top, \top]) \Rightarrow [\top, \top],
\end{aligned}
$$

therefore, by $(\mathcal{C} - 3)$ and $(\mathcal{C} - 10)$,

$$
\begin{aligned}
[\top, \top] &= [\top, \top] \Rightarrow [\top, \top] \\
&= ((Y \Rightarrow [\top, \top]) \Rightarrow [\top, \top]) \Rightarrow [\top, \top] \\
&= Y \Rightarrow [\top, \top],
\end{aligned}
$$

i.e., $Y \in \xi(\mathbb{A})$. This is a contradiction. Hence, $Y \twoheadrightarrow X \in \mathbb{A} - \xi(\mathbb{A})$ and $Y \Rightarrow X \in \mathbb{A} - \xi(\mathbb{A})$.

(ii) By the same method of part (i), suppose that $Y \twoheadrightarrow X \in \xi_d(\mathbb{A})$. Since $\xi_d(\mathbb{A})$ is subalgebra, $X \twoheadrightarrow (Y \twoheadrightarrow X) \in \xi_d(\mathbb{A})$. Thus, we have:

$$
\begin{aligned}
[\top, \top] &= (X \twoheadrightarrow (Y \twoheadrightarrow X)) \twoheadrightarrow [\top, \top] \\
&= (Y \twoheadrightarrow (X \twoheadrightarrow X)) \twoheadrightarrow [\top, \top] \\
&= (Y \twoheadrightarrow [\top, \top]) \twoheadrightarrow [\top, \top],
\end{aligned}
$$

therefore,

$$
\begin{aligned}
[\top, \top] &= [\top, \top] \twoheadrightarrow [\top, \top] \\
&= ((Y \twoheadrightarrow [\top, \top]) \twoheadrightarrow [\top, \top]) \twoheadrightarrow [\top, \top] \\
&= Y \twoheadrightarrow [\top, \top],
\end{aligned}
$$

i.e., $Y \in \xi_d(\mathbb{A})$. This is a contradiction. Also, if $Y \Rightarrow X \in \xi_d(\mathbb{A})$, since $\xi_d(\mathbb{A})$ is subalgebra, $X \Rightarrow (Y \Rightarrow X) \in \xi_d(\mathbb{A})$. Thus, we have:

$$
\begin{aligned}
[\top, \top] &= (X \Rightarrow (Y \Rightarrow X)) \Rightarrow [\top, \top] \\
&= (Y \Rightarrow (X \Rightarrow X)) \Rightarrow [\top, \top] \\
&= (Y \Rightarrow [\top, \top]) \Rightarrow [\top, \top],
\end{aligned}
$$

therefore, by $(\mathcal{C} - 3)$ and $(\mathcal{C} - 10)$,

$$
\begin{aligned}
[\top, \top] &= [\top, \top] \Rightarrow [\top, \top] \\
&= ((Y \Rightarrow [\top, \top]) \Rightarrow [\top, \top]) \Rightarrow [\top, \top]
\end{aligned}
$$

$$= Y \Rightarrow [\top, \top],$$

i.e., $Y \in \xi_d(\mathbb{A})$. This is a contradiction. Hence, $Y \twoheadrightarrow X \in U_d(\mathbb{A}) - \xi_d(\mathbb{A})$ and $Y \Rightarrow X \in U_d(\mathbb{A}) - \xi_d(\mathbb{A})$.

$\square$

Corollary 3.14. $\xi_d(\mathbb{A})$ is an interval ideal of $U_d(\mathbb{A})$.

Proof. It is an immediate consequence of Theorem 3.10,3.13. $\square$

4 Interval atoms in IBCI algebra

Definition 4.1. An element A of an IBCI algebra $\mathbb{A}$ is called an Interval-atom, if for every $X \in U_d(\mathbb{A})$, $A \precsim X$ or $A \ll X$, implies $X = A$.

The set of all interval atom elements of an IBCI algebra $\mathbb{A}$ is denoted by $\mathfrak{M}(\mathbb{A})$. It is straightforward $[\top, \top] \in \mathfrak{M}(\mathbb{A})$.

EXAMPLE 4. Let $\mathbb{A}$ be the IBCI algebra from Example 2. Then $\mathfrak{M}(\mathbb{A}) = U_d(\mathbb{A})$.

Theorem 4.2. *Let $\mathbb{A}$ be an IBCI algebra and $A \in U_d(\mathbb{A})$. If $A \in \mathfrak{M}(\mathbb{A})$, then for all $V_d \in U_d(\mathbb{A})$, $(A \twoheadrightarrow V_d) \twoheadrightarrow V_d = A$.*

Proof. Since $A \precsim (A \twoheadrightarrow V_d) \twoheadrightarrow V_d$ and $A \ll (A \twoheadrightarrow V_d) \twoheadrightarrow V_d$, by $(\mathcal{C}_d - 2)$, $(\mathcal{C}_d - 3)$, and A is an interval-atom, we have $(A \twoheadrightarrow V_d) \twoheadrightarrow V_d = A$.

$\square$

Theorem 4.3. *Let $\mathbb{A}$ be an IBCI algebra and $A \in \mathbb{A}$ satisfies the identity $X \twoheadrightarrow A = ((X \twoheadrightarrow A) \twoheadrightarrow Y) \twoheadrightarrow Y$ for all $X \in \mathbb{A}$. Then $A \in \mathfrak{M}(\mathbb{A})$.*

Proof. Assume that $Y \in \mathbb{A}$ such that $A \ll Y$ and $A \precsim Y$. Then

$$\begin{aligned} A = [\top, \top] &\twoheadrightarrow A \\ &= (([\top, \top] \twoheadrightarrow A) \twoheadrightarrow Y) \twoheadrightarrow Y \\ &= (A \twoheadrightarrow Y) \twoheadrightarrow Y = [\top, \top] \twoheadrightarrow Y = Y. \end{aligned}$$

$\square$

Theorem 4.4. *Let $\mathbb{A}$ be an IBCI algebra. If for all $V_d \in U_d(\mathbb{A})$, $(A \twoheadrightarrow V_d) \twoheadrightarrow V_d = A$, then $A \in \mathfrak{M}(\mathbb{A})$.*

Proof. Assume that for all $X \in U_d(\mathbb{A})$, $A \ll X$ and $A \precsim X$, then $A \rightarrow X = [\top, \top]$ and $A \Rightarrow X = [\top, \top]$, respectively. Therefore, we have

$$X = [\top, \top] \rightarrow X = (A \rightarrow X) \rightarrow X = A.$$

$\square$

Corollary 4.5. Let $\mathbb{A}$ be an IBCI algebra and $A \in U_d(\mathbb{A})$. Then $A \in U_d(\mathfrak{M}(\mathbb{A}))$ if and only if $A = (A \rightarrow [\top, \top]) \rightarrow [\top, \top]$.

Proof. Assume that $A \in U_d(\mathfrak{M}(\mathbb{A}))$. Since $A \ll (A \rightarrow [\top, \top]) \rightarrow [\top, \top]$, we have $A = (A \rightarrow [\top, \top]) \rightarrow [\top, \top]$. Conversely is hold, by putting $V_d = [\top, \top]$ in Theorem 4.4.

$\square$

Now, we can write $U_d(\mathfrak{M}(\mathbb{A})) = \{A \in U_d(\mathbb{A}) | A = (A \rightarrow [\top, \top]) \rightarrow [\top, \top]\}$. Therefore, since $A \Rightarrow [\top, \top] = A \rightarrow [\top, \top] = ((A \rightarrow [\top, \top]) \rightarrow [\top, \top]) \rightarrow [\top, \top]$, we can say that $A \rightarrow [\top, \top] = A \Rightarrow [\top, \top] \in U_d(\mathfrak{M}(\mathbb{A}))$.

Theorem 4.6. *Let $\mathbb{A}$ be an IBCI algebra. If for all $X \in \mathbb{A}$, $X \rightarrow A = (A \rightarrow X) \rightarrow [\top, \top]$. Then $A \in \mathfrak{M}(\mathbb{A})$.*

Proof. It's straightforward.

$\square$

Theorem 4.7. *Let $\mathbb{A}$ be an IBCI algebra. Then $A \in U_d(\mathfrak{M}(\mathbb{A}))$ if and only if for all $X, Y, A \in U_d(\mathbb{A})$, $X \rightarrow A = (A \rightarrow Y) \rightarrow (X \rightarrow Y)$.*

Proof. Assume that $A \in U_d(\mathfrak{M}(\mathbb{A}))$, then for all $X, Y \in U_d(\mathbb{A})$ and by Theorem 4.2, we have, $(A \rightarrow Y) \rightarrow (X \rightarrow Y) = X \rightarrow ((A \rightarrow Y) \rightarrow Y) = X \rightarrow A$.
Conversely, let $A, X \in U_d(\mathbb{A})$ and $A \ll X$, then $A \rightarrow X = [\top, \top]$. Now, Putting $Y = X$, we obtain that, by Proposition 2.4,

$$X \rightarrow A = (A \rightarrow X) \rightarrow (X \rightarrow X) = [\top, \top] \rightarrow [\top, \top] = [\top, \top],$$

i.e, $X \ll A$. Therefore $A \in U_d(\mathfrak{M}(\mathbb{A}))$.

$\square$

Lemma 4.8. *Let $\mathbb{A}$ be an IBCI algebra. If for all $A, X, Y \in \mathbb{A}$, $X \rightarrow A = (A \rightarrow Y) \rightarrow (X \rightarrow Y)$, then $X \Rightarrow A = ((X \Rightarrow A) \rightarrow Y) \Rightarrow Y$.*

Proof. For all $A, X, Y \in \mathbb{A}$ we have
$$\begin{aligned}((X \Rightarrow A) \rightarrow Y) \Rightarrow Y &= ((X \Rightarrow A) \rightarrow Y) \Rightarrow ([\top, \top] \rightarrow Y)\\ &= [\top, \top] \rightarrow (X \Rightarrow A)\\ &= X \Rightarrow A.\end{aligned}$$

$\square$

Corollary 4.9. Let $\mathbb{A}$ be an IBCI algebra. Then $A \in U_d(\mathfrak{M}(\mathbb{A}))$ if and only if for all $X, Y, A \in U_d(\mathbb{A})$, $X \Rightarrow A = ((X \Rightarrow A) \twoheadrightarrow Y) \Rightarrow Y$.

Proof. It follows from Theorem 4.7 and Lemma 4.8.

$\square$

Corollary 4.10. Let $\mathbb{A}$ be an IBCI algebra. Then for all $A, X \in U_d(\mathbb{A})$, the following statements are hold:

(i) $A \in U_d(\mathfrak{M}(\mathbb{A}))$.
(ii) $X \twoheadrightarrow A = (A \twoheadrightarrow [\top, \top]) \Rightarrow (X \twoheadrightarrow [\top, \top])$.
(iii) $X \Rightarrow A = (A \Rightarrow [\top, \top]) \twoheadrightarrow (X \Rightarrow [\top, \top])$.

Proof. It's clear by Theorem 4.7 and Proposition 2.3.

$\square$

Theorem 4.11. *Let $\mathbb{A}$ be an IBCI algebra. Then $U_d(\mathfrak{M}(\mathbb{A}))$ is an interval subalgebra.*

Proof. Assume that $X, Y \in U_d(\mathfrak{M}(\mathbb{A}))$, we need to show that $X \twoheadrightarrow Y, X \Rightarrow Y \in U_d(\mathfrak{M}(\mathbb{A}))$. Let $Z \in U_d(\mathbb{A})$, we have:

$$((X \twoheadrightarrow Y) \twoheadrightarrow [\top, \top]) \Rightarrow (Z \twoheadrightarrow [\top, \top])$$
$$= ((X \twoheadrightarrow [\top, \top]) \twoheadrightarrow (Y \twoheadrightarrow [\top, \top])) \Rightarrow (Z \twoheadrightarrow [\top, \top])$$
$$= (Y \twoheadrightarrow X) \Rightarrow (Z \twoheadrightarrow [\top, \top])$$
$$= (Y \Rightarrow X) \Rightarrow (Z \Rightarrow [\top, \top])$$
$$= Z \Rightarrow ((Y \Rightarrow X) \Rightarrow [\top, \top])$$
$$= Z \Rightarrow ((Y \twoheadrightarrow X) \twoheadrightarrow [\top, \top])$$
$$= Z \Rightarrow ((Y \twoheadrightarrow [\top, \top] \twoheadrightarrow (X \twoheadrightarrow [\top, \top])))$$
$$= Z \Rightarrow (X \twoheadrightarrow Y),$$

hence, by Corollary 4.10, $X \twoheadrightarrow Y \in U_d(\mathfrak{M}(\mathbb{A}))$ and by $(\mathcal{C} - 3)$, it is concluded $X \Rightarrow Y \in U_d(\mathfrak{M}(\mathbb{A}))$.

$\square$

Definition 4.12. An IBCI-algebra $\mathbb{A}$ is said to be p-semisimple if it satisfies for all $X \in \mathbb{A}$, if $[\top, \top] \lesssim X$, imply $X = [\top, \top]$.

Theorem 4.13. *Let $\mathbb{A}$ be an IBCI algebra. Then for all $X, Y \in U_d(\mathbb{A})$, the following statements are equivalent:*

(i) $\mathbb{A}$ *is p-semisimple,*

(ii) if $X \lesssim Y$, imply $X = Y$,

(iii) $(Y \Rightarrow X) \Rightarrow X = Y$,

(iv) $(X \Rightarrow [\top, \top]) \Rightarrow [\top, \top] = X$,

(v) $(Y \Rightarrow [\top, \top]) \Rightarrow X = (X \Rightarrow [\top, \top]) \Rightarrow Y$.

Proof. $(i) \rightarrow (ii)$: Let $X \lesssim Y$, then, by $(\mathcal{C} - 5)$ and Proposition 2.4, one has $Y \twoheadrightarrow X \lesssim X \twoheadrightarrow X$, i.e., $Y \twoheadrightarrow X \lesssim [\top, \top]$. Since $\mathbb{A}$ is p-semisimple, we have $Y \twoheadrightarrow X = [\top, \top]$ and by Proposition 2.4, $Y \Rightarrow X = [\top, \top]$, i.e., $X = Y$, by $(IBCI7)$.

$(ii) \rightarrow (iii)$: By $(\mathcal{C} - 2)$, $Y \lesssim (Y \Rightarrow X) \Rightarrow X$. Therefore, by (ii), we have $(Y \Rightarrow X) \Rightarrow X = Y$.

$(iii) \rightarrow (iv)$: By putting $Y = X$ and $X = [\top, \top]$ in (iii), we have $(X \Rightarrow [\top, \top]) \Rightarrow [\top, \top] = X$.

$(iv) \rightarrow (v)$: By (iv), we have
$$(Y \Rightarrow [\top, \top]) \Rightarrow X = (Y \Rightarrow [\top, \top]) \Rightarrow ((X \Rightarrow [\top, \top]) \Rightarrow [\top, \top])$$
$$= (X \Rightarrow [\top, \top]) \Rightarrow ((Y \Rightarrow [\top, \top]) \Rightarrow [\top, \top])$$
$$= (X \Rightarrow [\top, \top]) \Rightarrow Y.$$

$(v) \rightarrow (i)$: Let $X \lesssim [\top, \top]$. By $(IBCI4)$, and putting $Y = [\top, \top]$ in (v) we have:
$$X = [\top, \top] \twoheadrightarrow X = ([\top, \top] \twoheadrightarrow [\top, \top]) \twoheadrightarrow X$$
$$= (X \twoheadrightarrow [\top, \top]) \twoheadrightarrow [\top, \top]$$
$$= [\top, \top] \twoheadrightarrow [\top, \top] = [\top, \top],$$
i.e., $\mathbb{A}$ is p-semisimple. $\qquad\qquad\qquad\qquad\qquad\qquad\qquad\qquad\qquad\qquad\square$

Theorem 4.14. *Let $\mathbb{A}$ be an IBCI algebra. Then for all $X, Y, Z, W \in U_d(\mathbb{A})$, the following statements are equivalent:*

(i) $\mathbb{A}$ *is p-semisimple,*

(ii) $Y \twoheadrightarrow X = (X \twoheadrightarrow Y) \twoheadrightarrow [\top, \top]$,

(iii) $Y \twoheadrightarrow X = (X \twoheadrightarrow Z) \twoheadrightarrow (Y \twoheadrightarrow Z)$,

(iv) $Y \twoheadrightarrow X = (X \twoheadrightarrow [\top, \top]) \twoheadrightarrow (Y \twoheadrightarrow [\top, \top])$,

(v) *if* $Z \twoheadrightarrow X = W \twoheadrightarrow X$, *then* $Z = W$,

(vi) *if* $X \twoheadrightarrow Z = X \twoheadrightarrow W$, *then* $Z = W$.

Proof. $(i) \rightarrow (ii)$: By Theorem 4.13 and $(\mathcal{C} - 11)$, is clear.

$(ii) \rightarrow (i)$: By putting $Y = [\top, \top]$, we have $[\top, \top] \twoheadrightarrow X = (X \twoheadrightarrow [\top, \top]) \twoheadrightarrow [\top, \top]$. Hence, $X = (X \twoheadrightarrow [\top, \top]) \twoheadrightarrow [\top, \top]$. Now, by Theorem 4.13$(iv)$, $\mathbb{A}$ is p-semisimple.

$(i) \to (iii)$: By $(\mathcal{C} - 1)$ and Theorem 4.13(ii), property (iii) is holds.

$(iii) \to (i)$: Putting $Y = Z = [\top, \top]$ in (iii), we have

$$X = [\top, \top] \twoheadrightarrow X = (X \twoheadrightarrow [\top, \top]) \twoheadrightarrow ([\top, \top] \twoheadrightarrow [\top, \top])$$
$$= (X \twoheadrightarrow [\top, \top]) \twoheadrightarrow [\top, \top]$$
$$= (X \Rightarrow [\top, \top]) \Rightarrow [\top, \top],$$

now, by Theorem 4.13(iii), $\mathbb{A}$ is p-semisimple.

$(i) \leftrightarrow (iv)$: Since (i) and (iii) are equivalent, by putting $Z = [\top, \top]$ in (iii), we can say that (i) and (iv) are equivalent.

$(i) \to (v)$: Assume that $Z \twoheadrightarrow X = W \twoheadrightarrow X$. Since $\mathbb{A}$ is p-semisimple, we have

$$Z = (Z \twoheadrightarrow X) \twoheadrightarrow X = (W \twoheadrightarrow X) \twoheadrightarrow X = W.$$

$(v) \to (i)$: Let $X \lesssim Y$, for $X, Y \in U_d(\mathbb{A})$, then $X \Rightarrow Y = [\top, \top] = X \Rightarrow X$. Therefore, $X = Y$, by (v).

$(i) \to (vi)$: Let $X \twoheadrightarrow Z = X \twoheadrightarrow W$. Since $\mathbb{A}$ is p-semisimple, $Z \twoheadrightarrow X = (X \twoheadrightarrow Z) \twoheadrightarrow [\top, \top]$ and $W \twoheadrightarrow X = (X \twoheadrightarrow W) \twoheadrightarrow [\top, \top]$. Hence, by Theorem 4.13$(iii)$, we have

$$Z = (Z \twoheadrightarrow X) \twoheadrightarrow X = ((X \twoheadrightarrow Z) \twoheadrightarrow [\top, \top]) \twoheadrightarrow X$$
$$= ((X \twoheadrightarrow W) \twoheadrightarrow [\top, \top]) \twoheadrightarrow X$$
$$= (W \twoheadrightarrow X) \twoheadrightarrow X = W.$$

$(vi) \to (i)$: Let $X \lesssim Y$, where $X, Y \in U_d(\mathbb{A})$. Then $X \twoheadrightarrow Y = [\top, \top] = X \twoheadrightarrow X$, hence, by (vi), $X = Y$. Therefore, from Theorem 4.13(ii), $\mathbb{A}$ is p-semisimple. $\square$

5   Conclusion

In this paper the notion of interval ideals and atoms in IBCI-algebra was introduced and studied. We hope that this paper will encourage other researchers in this field to study other ideals and filters in BCI-algebra and find the relation between them.

References

[1] Bedregal, B. C. and Santiago, R. H. N. (2013). Interval representations, Łukasiewicz implicators and Smets-Magrez axioms. Information Sciences, 221:192–200.

[2] Bedregal, B. and Santiago, R. (2013). Some continuity notions for interval functions and representation. Computational and Applied Mathematics, 32(3):435–446.

[3] Cabrer, L. M. and Mundici, D. (2014). Interval $MV-$algebras and generalizations. International Journal of Approximate Reasoning, 55(8):1623–1642.

[4] Dudek, W. A. and Jun, Y. B. (2008). $Pseudo-$BCI algebras. East Asian Mathematical Journal, 24:187–190.

[5] Dymek, G. (2013). On the category of pseudo-BCI-algebras. Demonstratio Mathematica. Warsaw Technical University Institute of Mathematics, 46(4):631–644.

[6] Georgescu, G. and Iorgulescu, A. (2001). Pseudo-BCK algebras: An extension of BCK-algebras. In C. S. Calude, M. J. Dinneen, and S. Sburlan, editors, Combinatorics, Computability and Logic: Proceedings of the Third International Conference on Combinatorics, Computability and Logic, (DMTCS'01), pages 97–114, London. Springer London.

[7] Huang, Y. (2006). BCI-Algebra. Science Press, Beijing.

[8] Iorgulescu, A. (2016). New generalizations of BCI, BCK and Hilbert algebras. J. of Mult. Valued Logic and Soft Computing, Vol. 27, 4: 353–406.

[9] Iorgulescu, A. (2016). New generalizations of BCI, BCK and Hilbert algebras - Part II. J. of Mult.-Valued Logic and Soft Computing, Vol. 27, 4: 407–456.

[10] Iseki, K. (1966). An algebra related with a propositional calculus. Proc. Japan Acad., 42(1):26–29.

[11] Rasiowa, H. (1974). An Algebraic Approach to Non-classical Logics. Studies in Logic and the Foundations of Mathematics. North-Holland Publishing Company.

[12] Santiago, R. H. N., Bedregal, B. C. and Acioly, B. M. (2006). Formal aspects of correctness and optimality in interval computations. Formal Aspects of Computing, 18(2):231–243.

[13] Santiago, R. H. N., Bedregal, B. C., Marcos, J., Caleiro, C. and Pinheiro, J., (2019). Semi-BCI-Algebras. J. of Mult.-Valued Logic and Soft Computing, Vol. 32, pp. 87-109.

[14] Sunaga, T. (2009). Theory of an interval algebra and its application to numerical analysis [reprint of res. assoc. appl. geom. mem. 2 (1958), 29–46]. Japan J. Indust. Appl. Math., 26(2-3):125–143.

Received 5 October 2023

From Quasi-congruences to Congruences in Residuated Lattices

Farideh Farsad

Department of Pure Mathematics, Faculty of Mathematics and Computer, Higher Education Complex of Bam, Bam, Iran
faridehfarsad@yahoo.com

Arsham Borumand Saeid

Department of Pure Mathematics, Faculty of Mathematics and Computer, Shahid Bahonar University of Kerman, Kerman, Iran
arsham@uk.ac.ir

Mohammad Ali Nourollahi

Department of Pure Mathematics, Faculty of Mathematics and Computer, Higher Education Complex of Bam, Bam, Iran
mnourollahi@bam.ac.ir

Abstract

The isomorphism theorems describe the relationship between quotients, homomorphisms, and subobjects. Also, to construct the quotient of residuated lattices, whose algebraic structures is determined by a partial order, it is more useful to consider directed kernels of homomorphisms between such algebras. In this paper, we use the concept of directed kernel of a resituated lattice homomorphism, then introduce the concept of a quasi-congruence on a residuated lattice and then characterize them as the directed kernels of the residuated lattice homomorphisms. Finally, we prove the decomposition and isomorphism theorems for residuated lattice homomorphisms.

1   Introduction and Preliminaries

The interest in lattice-valued logic has been rapidly growing recently. Several algebras playing the role of structures of true values have been introduced and axiomatized [13].

Residuated lattices were first introduced as a generalization of ideal lattices of rings in 1939 by Ward and Dilworth[15].

For residuated lattices, whose algebraic structure is determined by a partial order, it is more useful to consider the directed kernel of residuated lattice homomorphisms to study the residuated lattice congruences. So, in this paper, we take a close look in directed kernels of residuated lattice homomorphisms and characterize them as quasi-congruences on a residuated lattices. Moreover, we characterize residuated lattice congruences as the intersection of a directed kernel relation and its inverse.

First of all, we recall some concepts of quasi-ordered sets which will be used in the sequel (see [2, 14]).

A congruence in an algebra is an equivalence which preserves all the algebraic operations. In every residuated lattice $(L, \wedge, \vee, \odot, \to)$, we show that an equivalence is a congruence iff it preserves both $\to$ and $\wedge$, iff it is respect to both $\to$ and $\vee$, for more information see [7].

Consider a set P with a reflexive and transitive relation σ. Such a relation will be called a *quasi-order* and (P, σ) *a quasi-ordered set*.

Let (P, ρ) be a quasi-ordered set. Then the relation $\overline{\rho}$ defined by $p \, \overline{\rho} \, q$ if and only if $(p, q) \in \rho \cap \rho^{-1}$, for each $p, q \in P$, is an equivalence relation on P. Moreover, it is the largest equivalence relation on P with the property that $\overline{\rho} \subseteq \rho$.

It is a well-known fact that $(P/\overline{\rho}, \rho/\overline{\rho})$ is a poset. Furthermore, we recall that $\overline{\rho}$ is the smallest equivalence relation with this property. Moreover, for every $X, Y \in P/\overline{\rho}$, $X \, (\rho/\overline{\rho}) \, Y$ if and only if $x \, \rho \, y$ for some $x \in X$ and $y \in Y$ if and only if $x \, \rho \, y$ for each $x \in X$ and $y \in Y$.

Recall that from[1, 5, 11], a bounded lattice is a structure $(L, \vee, \wedge, 0, 1)$ such that $(L, \vee, \wedge)$ is a lattice, 0 is the least element $0 \leq x$ and 1 is the greatest element $x \leq 1$.

A commutative ordered monoid for brevity commutative pomonoid is a commutative monoid M together with a partial ordering $\leq$ such that $0 \leq a$ for every $a \in M$, and $a \leq b$ implies $a + c \leq b + c$ for all $a, b, c \in M$.

Let's now recall for example from [15], the notion of residuated lattices and residuated lattice homomorphisms.

Definition 1.1. A *residuated lattice* is an algebra $(L, \wedge, \vee, \odot, \rightarrow, 0, 1)$ of type $(2, 2, 2, 2, 0, 0)$ equipped with an order $\leq$ satisfying the following:
(LR_1) $(L, \wedge, \vee, 0, 1)$ is a bounded lattice,
(LR_2) $(L, \odot, 1)$ is a commutative ordered monoid,
(LR_3) $\odot$ and $\rightarrow$ form an adjoint pair, i.e., $c \leq a \rightarrow b$ iff $a \odot c \leq b$, for all $a, b, c \in L$.

The relations between the pair of operations $\odot$ and $\rightarrow$ expressed by LR_3, is a particular case of the law of residuation, or Galois correspondence (see [11]) and for every $x, y \in A$, $x \rightarrow y = \sup\{z \in A \mid x \odot z \leq y\}$.

Definition 1.2. A function $f : L \rightarrow P$ between two residuated lattices L and P is called a *residuated lattice homomorphism* if it is a morphism of bounded lattices and for every $x, y \in L$:
$f(x \odot y) = f(x) \odot f(y)$ and $f(x \rightarrow y) = f(x) \rightarrow f(y)$.

Example 1.3. [14] Let $I = [0, 1]$. We define $\odot$ and $\rightarrow$ I as $x \odot y = \min\{x, y\}$ and

$$x \rightarrow y = \begin{cases} 1 & x \leq y \\ y & \text{otherwise} \end{cases}$$

for $x, y \in I$, is a residuated lattice.

Example 1.4. [14] Let $L = \{0, n, a, b, c, d, e, f, m, 1\}$ be a lattice ordered by the following figure

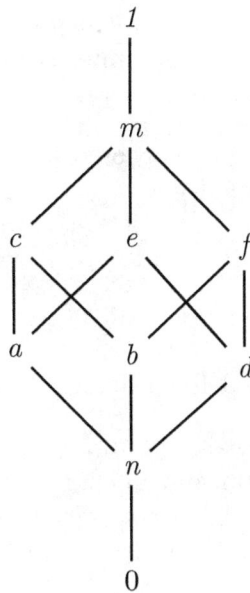

Define the operations $\rightarrow$ and $\odot$ by the following tables.

$\rightarrow$	0	n	a	b	c	d	e	f	m	1
0	1	1	1	1	1	1	1	1	1	1
n	m	1	1	1	1	1	1	1	1	1
a	f	f	1	f	1	f	1	f	1	1
b	e	e	e	1	1	e	e	1	1	1
c	d	d	e	f	1	d	e	f	1	1
d	c	c	c	c	c	1	1	1	1	1
e	b	b	c	b	c	f	1	f	1	1
f	a	a	a	c	c	e	e	1	1	1
m	n	n	a	b	c	d	e	f	1	1
1	0	n	a	b	c	d	e	f	m	1

$\odot$	0	n	a	b	c	d	e	f	m	1
0	0	0	0	0	0	0	0	0	0	0
n	0	0	0	0	0	0	0	0	0	n
a	0	0	a	0	a	0	a	0	a	a
b	0	0	0	b	b	0	0	b	b	b
c	0	0	a	b	c	0	a	b	c	c
d	0	0	0	0	0	d	d	d	d	d
e	0	0	a	0	a	d	e	d	e	e
f	0	0	0	b	b	d	d	f	f	f
m	0	0	a	b	c	d	e	f	m	m
1	0	n	a	b	c	d	e	f	m	1

Then L is a residuated lattice.

2 Quasi-complete quasi-ordered sets

Here, we introduce a generalization of complete lattices. Using this notion to define quasi-congruences on residuated lattices.

Definition 2.1. Let $(P, \leq)$ be a quasi-ordered set and $A \subseteq P$. An element $x \in P$ is called a *quasi-upper bound* of A if $a \leq x$, for all $a \in A$. A quasi-lower bound is defined dually. In other words, an element $y \in P$ is a *quasi-lower bound* of A if $y \leq a$, for all $a \in A$. The set of all quasi-upper bounds and quasi-lower bounds of A is denoted A^{PU} and A^{PL}, respectively.

Definition 2.2. Let $(P, \leq)$ be a quasi-ordered set and let $A \subseteq P$. An element $x \in P$ is said to be a *quasi-supremum* of A if

(i) x is a quasi-upper bound of A, and

(ii) $x \leq y$ for all quasi-upper bounds y of A.

A *quasi-infimum* of A is also defined dually.

Remark 2.3. Since $\leq$ is not necessarily anti-symmetric, a subset A of P may have more than one a quasi-supremum and quasi-infimum. But for every pair x, x' of quasi-suprema (quasi-infima), we have $x \leq x' \leq x$ or equivalently $x \overline{\leq} x'$. The set of all quasi-suprema and the set of all quasi-infima of A are denoted by Quasi-sup(A) and Quasi-inf(A), respectively.

In the following, we give an example of a quasi-ordered set which is not a poset. Also, we find quasi-supremum and quasi-infimum of it.

Example 2.4. Let X be a set with $|X| \geq 2$, and consider the total relation $\leq = X \times X$ on it. Then $(X, \leq)$ is a quasi-order. In fact, for every $A \subseteq X$, each element of X is a quasi-supremum and quasi-infimum of it. But $(X, \leq)$ is not a poset, because the binary relation $\leq$ is not a partial order.

Definition 2.5. Let $(P, \leq)$ be a non-empty quasi-ordered set.

(i) If Quasi-sup(A) $\neq \emptyset$, for all $A \subseteq P$, then P is called *pre-sup complete* or we say P has arbitrary quasi-suprema.

(ii) If Quasi-inf(A) $\neq \emptyset$, for all $A \subseteq P$, then P is called *quasi-inf complete* or we say P has arbitrary quasi-infima.

(iii) P is called *quasi-complete* if it is both quasi-sup complete and quasi-inf complete.

Lemma 2.6. *Let P be a non-empty quasi-ordered set such that $Quasi\text{-}sup(A) \neq \emptyset$, for all $A \subseteq P$. Then $Quasi\text{-}inf(A) \neq \emptyset$, for all $A \subseteq P$.*

Proof. Let $A \subseteq P$. By assumption Quasi-sup(A^{PL}) $\neq \emptyset$. We claim that Quasi-sup(A^{PL}) = Quasi-inf(A). First notice that every element of A is a quasi-upper bound of all quasi-infima of A. Hence every element $b \in$ Quasi-sup(A^{PL}) is a quasi-lower bound of A. Further, if c is any quasi-lower bound of A, then $c \in A^{PL}$ and so $c \leq b$ ($b \in$ Quasi-sup(A^{PL})). This implies that Quasi-sup(A') $\subseteq$ Quasi-inf(A). Conversely, if z is a quasi-infimum of A, then $z \in A^{PL}$ and also $t \leq z$ for all $z \in A^{PL}$. This shows that z is a quasi-upper bound of A'. Now, if w is any quasi-upper bound of A^{PL}, then $z \leq w$ ($z \in A^{PL}$). This implies that $z \in$ Quasi-sup(A^{PL}) and so Quasi-inf(A) $\subseteq$ Quasi-sup(A^{PL}), as required $\qquad \square$

Theorem 2.7. *Let P be a non-empty quasi-ordered set. Then the following are equivalent:*

(i) *P is quasi-complete.*

(ii) *$Quasi\text{-}sup(A) \neq \emptyset$, for all $A \subseteq P$.*

(iii) *$Quasi\text{-}inf(A) \neq \emptyset$, for all $A \subseteq P$.*

Proof. It is trivial that (i) implies (ii), and, by Lemma 2.6, (ii) implies (iii). By dually of the proof of the above lemma (iii) implies that (i). □

Definition 2.8. Let P and Q be two quasi-complete quasi-ordered sets and $f \colon P \to Q$ be a map.

(i) *f preserves arbitrary quasi-suprema* if s is a quasi-supremum of $A \subseteq P$ then $f(s)$ is a quasi-supremum of $f(A)$ in Q, for all $A \subseteq P$.

(ii) *f preserves arbitrary quasi-infima* if s is a quasi-infimum of $A \subseteq P$ then $f(s)$ is a quasi-infimum of $f(A)$ in Q, for all $A \subseteq P$.

3   Quasi-congruence relations on residuated lattices

In the following, we introduce the concept of a quasi-congruence relation on a residuated lattice which plays a significant role in characterizing residuated lattice congruences.

Definition 3.1. Let $(L, \leq)$ be a residuated lattice, a quasi-order σ on L is called *quasi-congruence* if it satisfies the following conditions:

(1) $\leq \subseteq \sigma$.

(2) (L, σ) has binary quasi-suprema and binary quasi-infima and $(L, \sigma, \odot)$ is a pomonoid.

(3) $\odot$ and $\to$ form an adjoint pair, i.e., $c \, \sigma \, (a \to b)$ if and only if $(a \odot c) \, \sigma \, b$, for all $a, b, c \in A$, where $a \to b = \text{quasi-suprema}\{c \in P \mid (a \odot c) \, \sigma \, b\}$.

(4) The identity function $id_L \colon (L, \leq) \longrightarrow (L, \sigma)$ preserves binary quasi-suprema and binary quasi-infima.

The following theorem gives a characterization of quasi-congruences that shows they are those relations which are compatible with operations on a residuated lattice.

Theorem 3.2. *Let $(L, \leq)$ be a residuated lattice and σ a quasi-order on P. Then σ is a quasi-congruence on L if and only if*

(i) $\leq \subseteq \sigma$;

(ii) If $a \, \sigma \, b$ and $a' \, \sigma \, b$ then $(a \vee a') \, \sigma \, b$, for all a, b, a' in L;

(iii) If $a \, \sigma \, b$ and $a' \, \sigma \, b$ then $(a \wedge a') \, \sigma \, b$, for all a, b, a' in L;

(iv) If $a \, \sigma \, b$ and $a' \, \sigma \, b$ then $(a \odot a') \, \sigma \, b$, for all a, b, a' in L;

(v) $\odot$ and $\rightarrow$ form an adjoint pair, i.e., $c \, \sigma \, (a \rightarrow b)$ iff $(a \odot c) \, \sigma \, b$, for all $a, b, c \in A$, where $a \rightarrow b = quasi\text{-}suprema\{c \in L \mid (a \odot c) \, \sigma \, b\}$.

Proof. Let σ be a quasi-congruence on L. Then condition (i) is exactly condition (1) in Definition 3.1. Using conditions (2) and (4) in Definition 3.1, we obtain that $a \vee b$ is a quasi-supremum and $a \wedge b$ is a quasi-infimum of finite subset $\{a, b\}$ in L. Hence, and by the definitions of quasi-suprema and quasi-infima, this implies conditions (ii) and (iii). In a same way, one can easily prove that conditions (ii) and (iii) also hold for $a \odot b$.

Conversely, by condition (i), $a \vee b$ is a quasi-upper bound of $A := \{a, b\}$ in (L, σ) and it is also a quasi-supremum of A by condition (ii), for each two-element subset A of L. This implies that Quasi-sup$(A) \neq \emptyset$ and also the identity map $id_L : (L, \leq) \longrightarrow (L, \sigma)$ preserves binary quasi-suprema. Moreover, condition (i) implies that $a \wedge b$ is a quasi-lower bound of A in (L, σ) and it is a quasi-infimum of A by condition (ii), for each two-elemet subset A of L. This means that Quasi-inf$(A) \neq \emptyset$ and the identity map $id_L : (L, \leq) \longrightarrow (L, \sigma)$ preserves the binary quasi-infima of each two-element subset A of L. Also, by condition (iii), it is obvious that id_L preserves $\odot$. Thus we have just proved the conditions (2) and (4) in the definition of a quasi-congruence, as required. $\qquad\qquad\qquad\qquad\qquad\qquad\qquad\qquad\qquad\qquad\qquad\qquad\qquad\qquad$ $\square$

Now, in the following we define the concept of a directed kernel for order-preserving maps which plays a crucial role in the theory of residuated lattice congruences.

Definition 3.3. Let $(P, \leq)$ and $(Q, \preceq)$ be two partially ordered sets and $f : P \rightarrow Q$ an order-preserving map. The set

$$\overrightarrow{ker} f := \{(a, b) \in P \times P \mid f(a) \preceq f(b)\},$$

is called the *directed kernel* of f (see [6]).

Example 3.4. Take the poset $\mathbb{N}^\infty = \mathbb{N} \cup \{\infty\}$ with the natural order, that is, $0 \sqsubseteq 1 \sqsubseteq 2 \sqsubseteq \ldots \sqsubseteq \infty$. Then $\mathbb{N}^\infty$ is a poset. Also, let $(P, \leq)$ be a poset such that $P = \{a_0, a_1, \ldots\} \cup \{b_0, b_1, \ldots\}$ and the order on P is $\leq = \{(a_i, b_i) : i \in \mathbb{N}\} \cup \{(a_{i+1}, b_i) : i \in \mathbb{N}\} \cup id_P$. Let $f : P \longrightarrow \mathbb{N}^\infty$ be the map defined by $f(a_{i+1}) = f(b_i) = i + 1$ for $i \in \mathbb{N}$, $f(a_0) = 0$. Trivially, f is a poset map. Moreover, $\overrightarrow{ker} f = \{(a_{i+1}, b_j) \mid j \geq i, \ i, j \in \mathbb{N}\} \cup \{(a_0, b_j) \mid j \in \mathbb{N}\} \cup \{(b_i, a_j) \mid j \geq i, \ j \neq 0, \ i, j \in \mathbb{N}\} \cup \{(a_i, a_j) \mid j \geq i, \ i, j \in \mathbb{N}\} \cup \{(b_i, b_j) \mid j \geq i, \ i, j \in \mathbb{N}\} \cup \mathrm{Id}_P$.

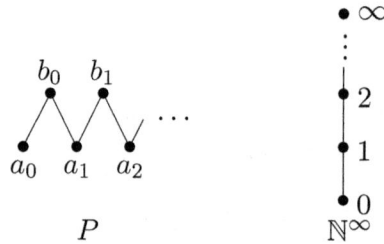

Theorem 3.5. *Let L and P be residuated lattices, and $f\colon L \longrightarrow P$ be a residuated lattice homomorphism. Then $\overrightarrow{\ker} f$ is a quasi-congruence on L.*

Proof. Since f is order-preserving, $\leq\,\subseteq\,\overrightarrow{\ker} f$. Now, let $a\,\overrightarrow{\ker} f\,b$ and $a'\,\overrightarrow{\ker} f\,b$ for all $a, b, a' \in L$. This gives that $f(a) \leq f(b)$ and $f(a') \leq f(b)$. So $f(a) \vee f(a') \leq f(b)$. Hence by $f(a \vee a') = f(a) \vee f(a')$, we have $f(a \vee a') \leq f(b)$. Thus $(a \vee a')\,\overrightarrow{\ker} f\,b$. Also by $f(a \wedge a') = f(a) \wedge f(a')$, we get $f(a \wedge a') \leq f(b \wedge b')$. This gives that $a \wedge a'\,\overrightarrow{\ker} f\,b$. Consequently, by Theorem 3.2, $\overrightarrow{\ker} f$ is a quasi-congruence. Also, if $a\,\overrightarrow{\ker} f\,(b \rightarrow c)$ then $f(a) \leq f(b \rightarrow c)$ and so $f(a) \leq f(b) \rightarrow f(c)$. By (LR3), one gets $f(a) \odot f(b) \leq f(c)$ then $f(a \odot b) \leq f(c)$, consequently, $a \odot b\,\overrightarrow{\ker}\,fc$. So we get to our result. $\qquad\qquad\square$

In the following we give a characterization of quasi-congruences as the directed kernels of residuated lattice homomorphisms.

Theorem 3.6. *Let $(L, \leq)$ be a residuated lattice and σ be a quasi-order on L. Then the following are equivalent:*

(1) σ is a quasi-congruence on L.

(2) $(L/\overline{\sigma}, \sigma/\overline{\sigma})$ is a residuated lattice, the canonical surjection

$$\pi\colon (L, \leq) \longrightarrow (L/\overline{\sigma}, \sigma/\overline{\sigma})$$

is a residuated lattice homomorphism and $\overrightarrow{\ker}\pi = \sigma$.

(3) There exist a residuated lattice $(P, \preceq)$ and a residuated lattice homomorphism $f\colon (L, \leq) \longrightarrow (P, \preceq)$ such that $\overrightarrow{\ker} f = \sigma$.

Proof. (1) $\Rightarrow$ (2): First we prove that $(L/\overline{\sigma}, \sigma/\overline{\sigma})$ has finite joins. To see this, let $A = \{[a_1], [a_2], \ldots, [a_n]\}$ be a finite subset of $L/\overline{\sigma}$. Then $t := \bigvee \pi_{\overline{\sigma}}^{-1}(A)$ exists in $(L, \leq)$ and so by hypothesis and condition (ii) in Theorem 3.2, t is a quasi-supremum of $\pi_{\overline{\sigma}}^{-1}(A)$ in (L, σ). Thus, since $\pi_{\overline{\sigma}}\colon (L, \sigma) \rightarrow (L/\overline{\sigma}, \sigma/\overline{\sigma})$ preserves and reflects the relations we obtain that $[t]$ is the supremum of $\pi_{\overline{\sigma}}(\pi_{\overline{\sigma}}^{-1}(A)) = A$ in

$(L/\bar{\sigma}, \sigma/\bar{\sigma})$, as required. Next, we prove that $(L/\bar{\sigma}, \sigma/\bar{\sigma})$ has finite meets. Take $s :=$ $\bigwedge\{a_1, a_2, \dots, a_n\}$ in $(L, \leq)$. So by condition (iii) in Theorem 3.2, s is a quasi-infimum of $\{a_1, a_2, \dots, a_n\}$ in (L, σ). Moreover, it is easy to show that Quasi-inf$(\pi_{\bar{\sigma}}^{-1}(A)) =$ Quasi-inf$(\{a_1, a_2, \dots, a_n\})$ in (L, σ). This gives that s is a quasi-infimum of $\pi_{\bar{\sigma}}^{-1}(A)$ in (L, σ). Now, since $\pi_{\bar{\sigma}} : (L, \sigma) \to (L/\bar{\sigma}, \sigma/\bar{\sigma})$ preserves and reflects the relations we obtain that $[s]$ is the infimum of $\pi_{\bar{\sigma}}(\pi_{\bar{\sigma}}^{-1}(A)) = A$ in $(L/\bar{\sigma}, \sigma/\bar{\sigma})$, as required. Since $(L, \leq)$ is a residuated lattice, so it is easy to show that $([a] \odot [b]) \sigma/\bar{\sigma} [c] \Leftrightarrow [a] \sigma/\bar{\sigma} ([b] \to [c])$ for $[a], [b]$ and $[c]$ in $L/\bar{\sigma}$. It is easy to check that $\pi_{\bar{\sigma}}$ is a homomorphism between residuated lattices. Finally,

$$\vec{\ker}\pi_{\bar{\sigma}} = \{(a, b) \mid \pi_{\bar{\sigma}}(a) \ \sigma/\bar{\sigma} \ \pi_{\bar{\sigma}}(b)\} = \{(a, b) \mid [a] \ \sigma/\bar{\sigma} \ [b]\}$$
$$= \{(a, b) \mid a \ \sigma \ b\} = \sigma.$$

$(2) \Rightarrow (3)$: Follows directly from the hypothesis.

$(3) \Rightarrow (1)$: It is followed immediately from Lemma 3.5. □

Now, we characterize residuated lattice congruences as the intersection of a quasi-congruence and its inverse.

Definition 3.7. An equivalence relation R on a residuated lattice $(L, \leq)$ is said to be *a residuated lattice congruence* if there is an order $\preceq$ on L/R such that $(L/R, \preceq)$ is a residuated lattice and the canonical map $\pi_R : L \to L/R$, $a \mapsto [a]_R$, is a residuated lattice map.

Theorem 3.8. *Let $(L, \leq)$ be a residuated lattice and R be an equivalence relation on L. Then the following are equivalent:*

(1) R is a residuated lattice congruence.

(2) There exists a quasi-congruence σ on L such that $R = \sigma \cap \sigma^{-1}$.

(3) There exist a residuated lattice P and a residuated lattice homomorphism $f : L \to P$ such that $R = \ker f$.

Proof. $(1) \Rightarrow (2)$: Let R be a residuated lattice congruence on L. Then there exists an order $\preceq$ on the quotient set L/R such that $(L/R, \preceq)$ is a residuated lattice and the canonical surjective map $\pi : L \to L/R$ is a residuated lattice homomorphism. Take $\sigma = \vec{\ker}\pi$. By Theorem 3.6, σ is a quasi-congruence on L. Moreover, $R = \ker \pi = \vec{\ker}\pi \cap (\vec{\ker}\pi)^{-1} = \bar{\sigma}$.

$(2) \Rightarrow (3)$: For a quasi-congruence σ on L, by Theorem 3.6, there exist a residuated lattice P and a surjective residuated lattice homomorphism $f : L \to P$ such that $\sigma = \vec{\ker}f$. Thus $R = \bar{\sigma} = \vec{\ker}f \cap (\vec{\ker}f)^{-1} = \ker f$.

$(3) \Rightarrow (1)$: By (3) and Theorem 3.6, $\sigma = \vec{\ker} f$ is a quasi-congruence on L. Also, $(L/R, \sigma/R)$ is a residuated lattice and the canonical surjective map $\pi\colon L \to L/R$ is a residuated lattice homomorphism, where $R = \bar{\sigma} = \ker f$. Thus, R is a residuated lattice congruence. $\qquad\square$

4 The Decomposition Theorem for residuated lattices

In this section, we prove Decomposition Theorem and its consequences for residuated lattices. We also prove a counterpart of algebra's first, second and third isomorphism theorems for residuated lattices.

Theorem 4.1 (The Decomposition Theorem). *Let $f\colon P \to Q$ and $g\colon P \to T$ be residuated lattice homomorphisms such that*

(1) *f is surjective,*

(2) *$\vec{\ker} f \subseteq \vec{\ker} g$.*

Then there is a unique residuated lattice homomorphism $h\colon Q \to T$ such that

$$h \circ f = g.$$

Moreover, h is injective if and only if $\vec{\ker} f = \vec{\ker} g$. Also, h is surjective if and only if g is surjective.

Proof. By Theorem 3.5, $\vec{\ker} f$ and $\vec{\ker} g$ are quasi-congruences. Theorem 3.6 implies that $\ker f$ and $\ker g$ are residuated lattice congruences. Also, by (2), $\vec{\ker} f \subseteq \vec{\ker} g$, and so

$$\ker f = (\vec{\ker} f) \cap (\vec{\ker} f)^{-1} \subseteq (\vec{\ker} g) \cap (\vec{\ker} g)^{-1} = \ker g.$$

Now, since f is surjective and $\ker f \subseteq \ker g$, by the Decomposition Theorem of functions, the map $h\colon Q \to T$ defined by $h(q) = g(p_q)$, where $p_q \in P$ is an element such that $f(p_q) = q$, is the unique function satisfying $h \circ f = g$. Moreover, h is surjective if and only if g is surjective.

Now, we show that h is a residuated lattice homomorphism. First notice that h preserves finite joins. To see this, let A be a finite subset of Q. Then, we show that $h(\bigvee A) = \bigvee h(A)$. Take $r := \bigvee A \in Q$ and $s := \bigvee f^{-1}(A) \in P$. So $f(s) = f(\bigvee f^{-1}(A)) = \bigvee f(f^{-1}(A)) = \bigvee A$ (notice that since f is surjective, $f(f^{-1}(A)) = A$). This gives that $h(\bigvee A) = g(s)$. We also have $h(A) = g(f^{-1}(A))$ and so $g(s) = g(\bigvee f^{-1}(A)) = \bigvee g(f^{-1}(A)) = \bigvee h(A)$. Consequently, $h(\bigvee A) = g(s) = \bigvee h(A)$, as required.

670

Now, we show that h preserves finite meets. To prove this, let B be a finite subset of Q. Take $z := \bigwedge B \in Q$ and $t := \bigvee f^{-1}(B) \in P$. So $f(t) = f(\bigwedge f^{-1}(B)) = \bigwedge f(f^{-1}(B)) = \bigwedge B$ (notice that $f(f^{-1}(B)) = B$, since f is surjective). Thus, $h(\bigwedge B) = g(t)$. We also have $h(B) = g(f^{-1}(B))$ and so $g(t) = g(\bigwedge f^{-1}(B)) = \bigwedge g(f^{-1}(B)) = \bigwedge h(B)$. Consequently, $h(\bigwedge B) = g(t) = \bigwedge h(B)$, as required.

Also, h preserves two operations $\odot$ and $\to$. In fact, since f is surjective so for $a, b \in Q$, we have $h(a) = h \circ f(f^{-1}(a)) = g(f^{-1}(a))$, then $h(a \odot b) = h \circ f(f^{-1}(a) \odot f^{-1}(b)) = g(f^{-1}(a) \odot f^{-1}(b)) = g(f^{-1}(a)) \odot g(f^{-1}(b)) = h(a) \odot h(b)$. By similar proof, we get $h(a \to b) = h(a) \to h(b)$.

Finally, suppose that $\ker f = \ker g$. We show that h is injective. To see this, let $h(q_1) = h(q_2)$ for $q_1, q_2 \in Q$. Since f is surjective, there exist $p_1, p_2 \in P$ such that $f(p_1) = q_1$, $f(p_2) = q_2$. Thus $g(p_1) = hf(p_1) = h(q_1) = h(q_2) = hf(p_2) = g(p_2)$. So, by hypothesis, $f(p_1) = f(p_2)$, that is $q_1 = q_2$.

Conversely, if $h \colon Q \to g(P)$ is injective, then $\vec{\ker} g = \vec{\ker} f$. By (2), $\vec{\ker} f \subseteq \vec{\ker} g$. To see $\vec{\ker} g \subseteq \vec{\ker} f$, let $x \vec{\ker} g\, y$. Then $g(x) \leq_T g(y)$ and so $g(x \wedge y) = g(x) \wedge g(y) = g(y)$. Thus $h(f(x \wedge y)) = g(x \wedge y) = g(y) = h(f(y))$. Therefore, $f(x) \wedge f(y) = f(x \wedge y) = f(y)$, since h is injective. So $f(x) \leq f(y)$. This gives that $x \vec{\ker} f\, y$, as required. $\qquad\square$

As a consequence, we have:

Corollary 4.2. *Let $g \colon P \to Q$ be a residuated lattice homomorphism. Also, let σ be a quasi-congruence on P that $\sigma \subseteq \vec{\ker} g$. Then there exists a unique residuated lattice homomorphism $h \colon P/\bar{\sigma} \to Q$ which makes the diagram*

$$
\begin{array}{ccc}
P & \xrightarrow{\ g\ } & Q \\
{\scriptstyle \pi}\big\downarrow & \nearrow{\scriptstyle h} & \\
P/\bar{\sigma} & &
\end{array}
$$

commutative. Moreover, h is injective if and only if $\sigma = \vec{\ker} g$. Also, h is surjective if and only if g is surjective.

Applying Theorem 3.8, we obtain the following corollary:

Corollary 4.3 (The First Isomorphism Theorem). *Let $f \colon P \to Q$ be a residuated lattice homomorphism. Then*

$$P/\ker f \cong f(P).$$

In particular, if $f \colon P \to Q$ is a surjective residuated lattice homomorphism, then

$$P/\ker f \cong Q.$$

Now, we give a description of quasi-congruences on the quotient residuated lattices.

Remark 4.4. (1) If $f\colon (P,\le) \to (Q,\sqsubseteq)$ is a surjective residuated lattice homomorphism and σ is a quasi-congruence on P such that $\overrightarrow{\ker}f \subseteq \sigma$, then $(f \times f)(\sigma)$ is a quasi-congruence on Q.

(2) Let $(P,\le)$ be a residuated lattice and σ, ρ two quasi-congruences on P with $\rho \subseteq \sigma$. Then the quotient relation

$$\sigma/\rho := \{([a]_{\bar\rho}, [b]_{\bar\rho}) \mid (a,b) \in \sigma\}$$

is a quasi-congruence on the residuated lattice $(P/\bar\rho, \rho/\bar\rho)$. This is because, by Theorem 3.6, the canonical surjection $\pi_{\bar\rho}\colon (P,\le) \to (P/\bar\rho, \rho/\bar\rho)$ is a residuated lattice homomorphism, and $\overrightarrow{\ker}\pi_{\bar\rho} = \rho \subseteq \sigma$. So, by (1), $\sigma/\bar\rho = (\pi_{\bar\rho} \times \pi_{\bar\rho})(\sigma)$ is a quasi-congruence on $(P/\bar\rho, \rho/\bar\rho)$.

(3) Every residuated lattice congruence of a quotient residuated lattice P/R is of the form $\overline{\theta/R}$, where $R = \bar\sigma$ (see Theorem 3.8), for some quasi-congruence σ on P, and θ is a quasi-congruence on P containing σ.

This is because by (2), for each pair of quasi-congruences θ and σ with $\sigma \subseteq \theta$, $\theta/\bar\sigma$ is a quasi-congruence on $P/\bar\sigma$. So, by Theorem 3.8, $\overline{\theta/\bar\sigma}$ is a residuated lattice congruences on P/R. On the other hand, for each residuated lattice congruence Θ on the quotient residuated lattice $(P/\bar\sigma, \sigma/\bar\sigma)$, by Theorem 3.8, $\Theta = \bar\rho$, for some quasi-congruence ρ on the residuated lattice $(P/\bar\sigma, \sigma/\bar\sigma)$. Now, $\theta := \{(x,y) \in P \times P \mid ([x]_{\bar\sigma}, [y]_{\bar\sigma}) \in \rho\}$ is a quasi-congruence on P containing σ, also $\rho = \theta/\bar\sigma$, and so $\Theta = \overline{\theta/\bar\sigma}$.

To prove the second isomorphism theorem for residuated lattices, we need the following lemma.

Lemma 4.5. *Let $(L,\le)$ be a residuated lattice and σ, ρ be two quasi-congruences on L with $\rho \subseteq \sigma$. Then $\text{Quasi-sup}_{(L,\rho)}(A) \subseteq \text{Quasi-sup}_{(L,\sigma)}(A)$ and also*

$$\text{Quasi-inf}_{(L,\rho)}(A) \subseteq \text{Quasi-inf}_{(L,\sigma)}(A),$$

for all $A \subseteq L$.

Proof. We have $\rho \subseteq \sigma$. This fact and $\bigvee A \in \text{Quasi-sup}_{(L,\rho)}(A) \cap \text{Quasi-sup}_{(L,\sigma)}(A)$ (see Theorem 3.6) imply that each quasi-supremum of every subset A of (L,ρ) is also a quasi-supremum of A in (Q,σ), as required. In the same way, one can easily show this fact for the quasi-infima. $\square$

Corollary 4.6 (The Second Isomorphism Theorem). *Let L be a residuated lattice, ρ, σ quasi-congruences on L with $\rho \subseteq \sigma$. Then*

$$L/\bar{\rho}\big/\overline{\sigma/\bar{\rho}} \cong L/\bar{\sigma}.$$

Proof. First, we show that $f\colon (L/\bar{\rho}, \rho/\bar{\rho}) \to (L/\bar{\sigma}, \sigma/\bar{\sigma})$, $[p]_{\bar{\rho}} \mapsto [p]_{\bar{\sigma}}$, is a residuated lattice homomorphism. First, we show that f preserves finite joins. To see this, let $A \subseteq (L/\bar{\rho}, \rho/\bar{\rho})$ and $[t]_{\bar{\rho}} := \bigvee A$ where t is a quasi-supremum of $\pi_{\bar{\rho}}^{-1}(A)$ in (L, ρ) (see the proof $(1) \Rightarrow (2)$ in Theorem 3.6). By Lemma 4.5, t is also a quasi-supremum of $\pi_{\bar{\rho}}^{-1}(A)$ in (L, σ). Hence $[t]_{\bar{\sigma}} = f([t]_{\bar{\rho}})$ is a supremum of

$$
\begin{aligned}
\{[x]_{\bar{\sigma}} \mid x \in \pi_{\bar{\rho}}^{-1}(A)\} &= \{[x]_{\bar{\sigma}} \mid [x]_{\bar{\rho}} \in A\} \\
&= \{f([x]_{\bar{\rho}}) \mid [x]_{\bar{\rho}} \in A\} \\
&= f(A).
\end{aligned}
$$

Hence f preserves finite joins. In the same way, one can show that f preserves finite meets. Also, it is easy to check that f preserves $\to$ and $\odot$ and conclude that f is a residuated lattice homomorphism.

Finally, by the definition of $\overrightarrow{\ker} f$ and the quotient relations $\sigma/\bar{\sigma}$ and $\sigma/\bar{\rho}$, we have

$$
\begin{aligned}
([a]_{\bar{\rho}}, [b]_{\bar{\rho}}) \in \overrightarrow{\ker} f &\Leftrightarrow [a]_{\bar{\sigma}}\, \sigma/\bar{\sigma}\, [b]_{\bar{\sigma}} \\
&\Leftrightarrow (a, b) \in \sigma \\
&\Leftrightarrow ([a]_{\bar{\rho}}, [b]_{\bar{\rho}}) \in \sigma/\bar{\rho}.
\end{aligned}
$$

So, $\ker f = \overrightarrow{\ker} f \cap (\overrightarrow{\ker} f)^{-1} = \sigma/\bar{\rho} \cap (\sigma/\bar{\rho})^{-1} = \overline{\sigma/\bar{\rho}}$. Now, by Corollary 4.3, we have $L/\bar{\rho}\big/\overline{\sigma/\bar{\rho}} \cong L/\bar{\sigma}$. $\qquad\square$

To prove the third isomorphism theorem, we study the relation between the quasi-congruences on a residuated lattice and its subresiduated lattices.

Lemma 4.7. *Let $(L, \leq)$ be a residuated lattice, M a subresiduated lattice of L and σ a quasi-congruence on $(L, \leq)$. Then $\sigma \upharpoonright_M = \sigma \cap (M \times M)$ is a quasi-congruence on M and $\overline{\sigma \upharpoonright_M} = \bar{\sigma} \cap (M \times M)$.*

Proof. By Theorem 3.2, since σ is a quasi-congruence on L and M is closed under arbitrary joins and finite meets, $\sigma \upharpoonright_M$ is trivially a quasi-congruence on M. Moreover,

$$
\begin{aligned}
\overline{\sigma \upharpoonright_M} &= \sigma \upharpoonright_M \cap (\sigma \upharpoonright_M)^{-1} \\
&= (\sigma \cap (M \times M)) \cap (\sigma \cap (M \times M))^{-1} \\
&= (\sigma \cap (M \times M)) \cap (\sigma^{-1} \cap (M \times M)) \\
&= (\sigma \cap \sigma^{-1}) \cap (M \times M) \\
&= \bar{\sigma} \cap (M \times M).
\end{aligned}
$$

□

Lemma 4.8. *Let* $(L, \leq)$ *be a residuated lattice,* M *a subresiduated lattice of* L *and* σ *a quasi-congruence on* $(L, \leq)$*. Then* $M^\sigma := \{a \in L \mid M \cap [a]_{\bar{\sigma}} \neq \emptyset\}$ *is a subresiduated lattice of* L*.*

Proof. First notice that LR_3 holds in any subset of L. Hence it is enough to show that M^σ is closed under its binary operations. To see this, first let A be an finite subset of M^σ. Then for all $a \in A$ there exists $q_a \in M \cap [a]_{\bar{\sigma}}$. Take $S := \{q_a \mid a \in A\}$ and $s := \bigvee S$. So $a \, \sigma \, q_a \leq s$, for all $a \in A$. This gives that $a \, \sigma \, s$, for all $a \in A$ (notice that $\leq \subseteq \sigma$ and σ is transitive). Hence and by Theorem 3.2, $\bigvee A \, \sigma \, s$. Furthermore, $q_a \, \sigma \, a \leq \bigvee A$, for all $a \in A$. This gives that $q_a \, \sigma \, \bigvee A$, for all $q_a \in S$. Hence and by Theorem 3.2, $s = (\bigvee S) \, \sigma \, (\bigvee A)$. Consequently, $(\bigvee A) \, \sigma \, s \, \sigma \, (\bigvee A)\}$ implies that $s \in [\bigvee A]_{\bar{\sigma}}$ and so $[\bigvee A]_{\bar{\sigma}} \cap M \neq \emptyset$ (notice that M is a subresiduated lattice and $S \subseteq M$). Hence $\bigvee A \in M^\sigma$, as required.

Second, let B be a finite subset of M^σ. Then for all $b \in B$ there exists $q_b \in M \cap [b]_{\bar{\sigma}}$. Take $T := \{q_b \mid b \in B\}$ and $t := \bigwedge T$. So $t \leq q_b \, \sigma \, b$, for all $b \in B$. This gives that $t \, \sigma \, b$, for all $b \in B$ (notice that $\leq \subseteq \sigma$ and σ is transitive). Hence and by Theorem 3.2, $t \, \sigma \, \bigwedge B$. Furthermore, $\bigwedge B \leq b \, \sigma \, q_b$, for all $b \in B$. This gives that $\bigwedge B \, \sigma \, q_b$, for all $q_b \in T$. Hence by Theorem 3.2, $t = (\bigwedge T) \, \sigma \, (\bigwedge B)$. Consequently, $(\bigwedge B) \, \sigma \, t \, \sigma \, (\bigwedge B)\}$ implies that $t \in [\bigwedge B]_{\bar{\sigma}}$ and so $[\bigwedge B]_{\bar{\sigma}} \cap M \neq \emptyset$ (notice that M is a subresiduated lattice and $S \subseteq M$). Hence $\bigwedge B \in M^\sigma$, as required.

Third, let $a, b \in M^\sigma$, then there exist $q_a \in M \cap [a]_{\bar{\sigma}}$ and $q_b \in M \cap [b]_{\bar{\sigma}}$. So $q_a \odot q_b \in M$ and also $q_a \, \sigma \, a$, $a \, \sigma \, q_a$ and $b \, \sigma \, q_b$, $q_b \, \sigma \, b$. By LR_2, we have $q_a \odot q_b \, \sigma \, a \odot b$ and $a \odot b \, \sigma \, q_a \odot q_b$. Therefore $q_a \odot q_b \in [a \odot b]_{\bar{\sigma}}$, this means that $[a \odot b]_{\bar{\sigma}} \cap M \neq \emptyset$ or $a \odot b \in M^\sigma$ as we needed.

Finally, we have $q_a \to q_b \in M$ and $a \, \sigma \, q_a$. Then by LR_2

$$a \odot (q_a \to q_b) \, \sigma \, q_a \odot (q_a \to q_b) \, \sigma \, q_b \, \sigma \, b.$$

Hence $a \odot (q_a \to q_b) \, \sigma \, b$, so $q_a \to q_b \, \sigma \, a \to b$ by LR_3. On the other hand, from $q_a \, \sigma \, a$ we get

$$q_a \odot (a \to b) \, \sigma \, a \odot (a \to b) \, \sigma \, b \, \sigma \, q_b.$$

Then $q_a \odot a \to b \, \sigma \, q_b$, so $a \to b \, \sigma \, q_a \to q_b$. Therefore $q_a \to q_b \in [a \to b]_{\bar{\sigma}}$, this means that $[a \to b]_{\bar{\sigma}} \cap M \neq \emptyset$ or $a \to b \in M^\sigma$ as we needed. Consequently, M^σ is a subresiduated lattice of L. □

Theorem 4.9 (The Third Isomorphism Theorem). *Let* $(L, \leq)$ *be a residuated lattice,* M *a subresiduated lattice of* L *and* σ *be a quasi-congruence on* L*. Then*

$$M / \overline{\sigma \restriction_M} \cong M^\sigma / \overline{\sigma \restriction_{M^\sigma}}$$

674

Proof. We consider the residuated lattice homomorphism $\pi_{\overline{\sigma\restriction M^\sigma}} \circ i\colon M \to M^\sigma/\overline{\sigma\restriction M^\sigma}$, where $i\colon M \to M^\sigma$ is the inclusion map. Also, we have $\overrightarrow{\ker}\pi_{\overline{\sigma\restriction M^\sigma}} \circ i = \sigma\restriction_M$. In fact,

$$
\begin{aligned}
(m, m') \in \overrightarrow{\ker}\pi_{\overline{\sigma\restriction M}} \circ i \;\;&\Leftrightarrow\;\; \pi_{\overline{\sigma\restriction M}} \circ i(m) \le \pi_{\overline{\sigma\restriction M}} \circ i(m') \\
&\Leftrightarrow\;\; \pi_{\overline{\sigma\restriction M}}(m) \le \pi_{\overline{\sigma\restriction M}}(q') \\
&\Leftrightarrow\;\; (m, m') \in (\overrightarrow{\ker}\pi_{\overline{\sigma\restriction M}}) \cap (M \times M) \\
&\Leftrightarrow\;\; (m, m') \in (\sigma\restriction_{M^\sigma}) \cap (M \times M) \\
&\Leftrightarrow\;\; (m, m') \in (\sigma \cap (M^\sigma \times M^\sigma)) \cap (M \times M) \\
&\Leftrightarrow\;\; (m, m') \in \sigma \cap ((M^\sigma \times M^\sigma) \cap (M \times M)) \\
&\Leftrightarrow\;\; (m, m') \in \sigma \cap (M \times M) = \sigma\restriction_M
\end{aligned}
$$

So, By Corollary 4.2, there is a unique injective residuated lattice homomorphism $h\colon M/\overline{\sigma\restriction M} \to M^\sigma/\overline{\sigma\restriction M^\sigma}$ such that the diagram is

commutative. In other words, h given by $h([m]_{\overline{\sigma\restriction M}}) = [m]_{\overline{\sigma\restriction M^\sigma}}$. To finish the proof, it is enough to show that h is also surjective. By the definition of M^σ, for all $a \in M^\sigma$ there exists a $m_a \in M$ with $m_a \in [a]_{\bar\sigma}$. Thus $h([m_a]_{\overline{\sigma\restriction M}}) = [m_a]_{\overline{\sigma\restriction M^\sigma}} = [a]_{\overline{\sigma\restriction M^\sigma}}$, as required (notice that by Lemma 4.7, $\overline{\sigma\restriction}_M = \bar\sigma \cap (M \times M)$ and $\overline{\sigma\restriction}_{M^\sigma} = \bar\sigma \cap (M^\sigma \times M^\sigma)$). $\qquad\square$

5 Conclusions

In this paper, we looked at the quotients of residuated lattices in a different way from that ones already exist. In fact, in set theory we can consider any set with the identity order Δ and then any function between sets is indead an order-preserving map. In this point of view, the kernel of a function $f : (X, =) \to (Y, =)$ is defined as the set of ordered pairs (a, b) of elements of its domain for which $f(a) = f(b)$. By generalizing this consept to sets with any arbitrary orders, namely posets, we introduced the concept of a directed kernel for an order-preserving map $f : (P, \leq) \to (Q, \sqsubseteq)$ between posets, that is, the set of ordered pairs (a, b) of elements of P with $f(a) \sqsubseteq f(b)$. Having this idea, we introduced the concept of a quasi-congruence on a residuated lattice and charecterized them as the intersection of a directed kernel of a residuated lattice homomorphism and its inverse. Using this characterization, we proved that the quotients of residuated lattices are the intersection of such quasi-congruences and their inverse.

Every function $f : A \to B$ can be factored through its image, that is, written as a composite $f = m \circ e$, where $e : A \to f(A)$ is a codomain resteriction of A and $m : f(A) \to B$ is the inclusion. In other words, every function has a factorization $f = m \circ e$ where m is onto and e is one-one. So, in this paper factorization structures for resituated lattice homorphisms are investigated. Having this idea, we also proved the counterpart of algebras first, second and third isomorphism theorems for residuated lattices.

The authors suggest these notions for other ordered algebras.

References

[1] Birkhoff, G., "Lattice Theory", 3rd ed., AMS Colloquium Publications, Vol. 25, 1967.

[2] Blyth, T.S., Janowitz, M.F., "Residuation Theory", International Series of Monographs in Pure and Applied Mathematics, Vol. 102, Pergamon Press, Oxford-New York-Toronto, Ont., 1972.

[3] Burris, S., Sankapannavar, H. P., "A Course in Universal Algebra", Springer-Verlag, 1981.

[4] Esteva, F., Godo L., *Monoidal tnorm based logic: towards a logic for left-continnuos tnorms, Fuzzy Sets and Systems*, (2001), 124(3), 271-288.

[5] Davey, B., Priestley. H., "Introduction to Lattices and Order", 2nd ed, Cambridge University Press, 2002.

[6] Farsad, F., Moghbeli-Damaneh, H., *Weighted limits in the category* **Dcpo**-*S*, Asian-European Journal of Mathematics 10(2) (2017) 1750037 (19 pages).

[7] Feng, S., Yang, J., *Congruences in residuated lattices*, 7th International Conference on Modelling, Identification and Control, (2015), 18-20.

[8] Galatos, N., Jipsen, P., Tomasz Kowalski and Hiroakira Ono, Residuated Lattices: an algebraic glimpse at substructural logics, Studies in Logics and the Foundations of Mathematics, Elsevier, 2007.

[9] Hajek P., *Mathematics of Fuzzy Logic*, (1998), *Dordrecht: Kluwer Academic Publshers.*

[10] Iorgulescu A., *Algebras of logic as BCK algebra*, 2008, Romania: *Academy of Economic Studies Bucharest.*

[11] Johnstone, P.T., " Stone Space", Cambridge University Press 1982.

[12] Kondo M., Modal operators on commutative residuated lattices, *Math. Slovaca*, 2011, 61, 1-14.

[13] Kowalski T., Ono H., Residuated lattices: An algebraic glimpse at logics without contraction, *JAIST*, 2002, 19-56.

[14] Piciu D., Algebras of Fuzzy Logic, 2007, Craiova: *Editura Universitaria Craiova.*

[15] Ward M., Dilworth R. P., *Residuated lattices*, *Trans. Am. Math. Soc.*, (1939), 45, 335-354.

Received 12 October 2023